RENEWABLE ENERGY: RESEARCH, DEVELOPMENT AND POLICIES

VOLUNTARY MARKETS IN RENEWABLE ENERGY

STATUS, TRENDS AND OPTIONS

RENEWABLE ENERGY: RESEARCH, DEVELOPMENT AND POLICIES

Additional books in this series can be found on Nova's website under the Series tab.

Additional E-books in this series can be found on Nova's website under the E-book tab.

ENERGY POLICIES, POLITICS AND PRICES

Additional books in this series can be found on Nova's website under the Series tab.

Additional E-books in this series can be found on Nova's website under the E-book tab.

RENEWABLE ENERGY: RESEARCH, DEVELOPMENT AND POLICIES

VOLUNTARY MARKETS IN RENEWABLE ENERGY

STATUS, TRENDS AND OPTIONS

JAMIR S. FULLER
EDITOR

nova publishers
New York

For permission to use material from this book please contact us:
Telephone 631-231-7269; Fax 631-231-8175
Web Site: http://www.novapublishers.com

Additional color graphics may be available in the e-book version of this book.

LIBRARY OF CONGRESS CATALOGING-IN-PUBLICATION DATA

ISBN: 978-1-62417-561-9

Published by Nova Science Publishers, Inc. † New York

CONTENTS

PREFACE

This book examines 2011 trends in the U.S. "voluntary" markets - those in which consumers and institutions purchase renewable energy to match their electricity needs on a voluntary basis. This review uses available data sources to capture 2011 voluntary market trends, status and options.

Chapter 1 - This report documents 2011 trends in the U.S. “voluntary” markets – those in which consumers and institutions purchase renewable energy to match their electricity needs on a voluntary basis. Although a full estimate of the size of the voluntary market is not available for 2011, this review uses available data sources to capture 2011 voluntary market trends.

Chapter 2 – This guide explores five innovative options for voluntarily procuring renewable energy generation or systems. These methods can be replicated by a variety of stakeholders—including local governments, not-for-profit organizations, businesses, and utilities. The options described in this guide offer a variety of benefits, including lower cost, simplified transactions, expanded access to green energy, and increased impact. Each chapter describes one option, provides examples of its use, lays out design and implementation considerations, and discusses benefits and challenges.

Chapter 3 – Voluntary markets for renewable energy in which consumers purchase renewable energy from their electricity providers or from renewable energy certificate (REC) marketers have existed for more than a decade. Going forward, various factors will influence the development of these markets, including potential climate policies, state and federal renewable portfolio standards (RPS), renewable energy prices, the level of consumer interest in purchasing green power, and the interest on the part of utilities in offering new green options and in continuing to promote existing programs.

This report presents estimates of voluntary market demand for renewable energy or green power through 2015. The forecast relies on historical data, including market sector sizes and growth rates, to inform our assumptions. Though we adjusted growth rates as discussed in the report, they remain the starting point for our analysis. Because of the different factors that affect utility programs, REC markets, and competitive electric markets, separate assumptions and methods are used to project demand for each submarket.

Chapter 4 – Voluntary markets for renewable energy in which consumers purchase renewable energy from their electricity providers or from renewable energy certificate (REC) marketers have existed for more than a decade. Going forward, various factors will influence the development of these markets, including potential climate policies, state and federal renewable portfolio standards (RPS), renewable energy prices, the level of consumer interest in purchasing green power, and the interest on the part of utilities in offering new green options and in continuing to promote existing programs.

This report presents estimates of voluntary market demand for renewable energy or green power through 2015. The forecast relies on historical data, including market sector sizes and growth rates, to inform our assumptions. Though we adjusted growth rates as discussed in the report, they remain the starting point for our analysis. Because of the different factors that affect utility programs, REC markets, and competitive electric markets, separate assumptions and methods are used to project demand for each submarket.

In: Voluntary Markets in Renewable Energy
Editor: Jamir S. Fuller
ISBN: 978-1-62417-561-9

Chapter 1

MARKET BRIEF: STATUS OF THE VOLUNTARY RENEWABLE ENERGY CERTIFICATE MARKET (2011 DATA)*

Jenny Heeter, Philip Armstrong and Lori Bird

LIST OF ACRONYMS

CEC	Clean Energy Collective
CRS	Center for Resource Solutions
DOE	U.S. Department of Energy
EERE	Office of Energy Efficiency and Renewable Energy
EIA	Energy Information Administration
EPA	Environmental Protection Agency
ERCOT	Electric Reliability Council of Texas
IOU	investor-owned utility
ISO	independent system operator
kW	kilowatt
kWh	kilowatt-hour
MI-RECS	Michigan Renewable Energy Certification System
M-RETS	Midwest Renewable Energy Tracking System

* This is an edited, reformatted and augmented version of the National Renewable Energy Laboratory, NREL/TP-6A20-56128, dated September 2012.

MW	megawatt
MWh	megawatt-hour
NAR	North American Renewables Registry
NC-RETS	North Carolina Renewable Energy Tracking System
NEPOOL-GIS	New England Power Pool-Generation Information System
NREL	National Renewable Energy Laboratory
NVTREC	Nevada Tracks Renewable Energy Credits
NYSERDA	New York State Energy Research and Development Authority
PJM-GATS	PJM-Generation Attribute Tracking System
PUC	public utility commission
REC	renewable energy certificate
RPS	renewable portfolio standard
RTO	regional transmission organization
SMUD	Sacramento Municipal Utility District
WECC	Western Electricity Coordinating Council
WREGIS	Western Renewable Energy Generation Information System

EXECUTIVE SUMMARY

This report documents 2011 trends in the U.S. “voluntary” markets – those in which consumers and institutions purchase renewable energy to match their electricity needs on a voluntary basis. Although a full estimate of the size of the voluntary market is not available for 2011, this review uses available data sources to capture 2011 voluntary market trends. Based on this review, the following key trends have been identified:

- In 2011, indicative data show that voluntary sales of renewable energy continued to increase (Figure ES-1).
  A comprehensive estimate for 2011 is not available, but several data sources provide indications of market trends. Regional or state renewable energy certificate (REC) tracking systems provide a partial estimate of the market because not all trasactions are captured in these systems.
  In 2011, voluntary market sales captured in REC tracking systems totaled approximately 40 million MWh, up from 20.8 million in

2010—a growth rate of 90%. However, a large portion of this growth likely represents an increase in the use of tracking systems for verifying voluntary market transactions, rather than an increase in purchases in the marketplace. For comparison, NREL estimated the voluntary market at 35.6 million MWh in 2010 based on data provided by market participants.

- Data from other available sources representing portions of the voluntary market show growth rates of approximately 20% for 2011. For example, the Environmental Protection Agency's (EPA) Green Power Partnership grew to 21.7 million MWh in 2011, representing growth of 22% from 2010.
  The EPA data capture only purchases of corporate and institutional purchases who are members of the EPA's program; however, EPA data does capture nearly all of the largest corporate purchases.
- The Center for Resource Solutions' Green-e Energy program has a preliminary estimate for 2011 showing a total of 27.9 million MWh in 2011, representing growth of 21% from 2010. As not all market transactions are Green-e Energy certified, these data represent a subset of the market. In 2010, NREL estimated that Green-e Energy retail transactions represented about 65% of the total market volume.
- On a MWh basis, the Electric Reliability Council of Texas (ERCOT) and the Western Renewable Energy Generation Information System (WREGIS) saw the largest growth in 2011. Voluntary retirements in ERCOT increased by 29% between 2010 and 2011, from 11.8 million MWh in 2010 to 15.3 million MWh in 2010.[1] In 2011, an additional 4.9 million MWh generated in 2009 and 2010 were also retired, for a total of 20.2 million MWh of voluntary retirements in 2011. Voluntary retirements in WREGIS increased to 8.3 million MWh in 2011 from 1.1 million MWh in 2010.
- Based on data now available from EIA, the largest 20 programs that report in the monthly EIA 826 form saw mixed sales results in 2011. A few programs saw sales growth greater than 35%, while others saw losses of similar proportion; the median growth for the sample of programs was 1%.
- Community solar programs have been growing, with 10 programs added in 2011. As of August 2012, nearly 30 community solar programs were operating, totaling 10 MW of combined capacity.

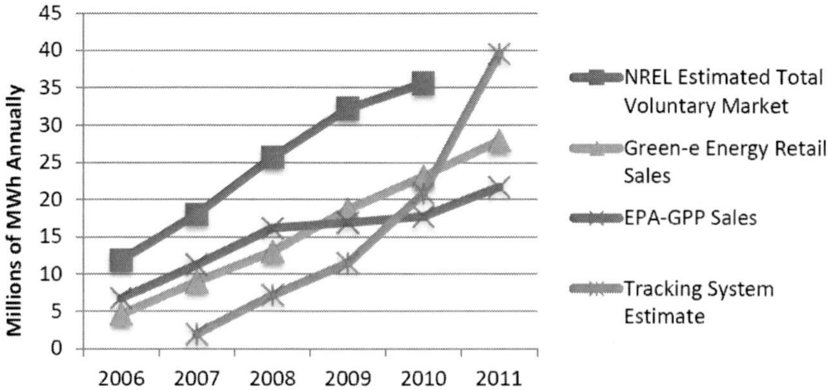

Sources: EIA 2012, EPA 2012a, Heeter and Bird 2011, Terada 2012.

Figure ES-1. Estimated annual voluntary sales by market sector, 2006–2011.

1. Introduction

Voluntary markets for renewable energy, or "green power" markets, allow consumers and institutions to purchase renewable energy to match their electricity needs on a voluntary basis. Entities can make voluntary purchases of renewable energy through utility green power programs and green power marketing activities in competitive electricity markets, as well as in unbundled REC markets. This report covers all of these approaches:

- *Utility Green Pricing (regulated utility markets).* Utility green pricing programs began in the early 1990s when a small number of utilities offered their customers the option to purchase green power. These programs are offered by utilities in traditionally regulated electricity markets. More than 850 utilities offer green pricing programs.
- *Competitive Green Power (competitive utility markets).* In states with competitive (or restructured) retail electricity markets, electricity customers may have the option of switching to an alternative electricity supplier that offers green power. In some of these states, default utility electricity suppliers offer green power options to their customers in conjunction with competitive green power marketers so that switching is not required. More than a dozen states that have opened their markets to retail competition have experienced some green power marketing activity.[2]

- *Voluntary Unbundled REC Market.* Regardless of whether customers have access to a green power product from their retail power provider, they can purchase green power through unbundled RECs. More than 25 companies offer unbundled RECs to retail customers via the Internet, and a number of other companies market RECs solely to commercial and wholesale customers.[3]

This analysis reviews 2011 trends in the voluntary renewable energy market based on available data. For more detailed background information, 2010 market data, and historical analysis, see Heeter and Bird 2011. Unlike in previous NREL reports,[4] we are not able to provide a comprehensive estimate of the market size because we did not collect data from market participants. Instead, we rely on publicly available data sources, which provide an indication of market performance for various portions of the market. However, these estimates cannot be compared to earlier NREL estimates of the size of the voluntary market, because of the differences in methodology and data sources.

This analysis draws on data from four sources that capture portions of the voluntary market. These data overlap, and are presented separately in order to examine growth trends, but cannot be summed to create a complete picture of the voluntary market in 2011. Each data source covers only a subsection of the voluntary market.

Data sources used in this report are as follows:

1. The **U.S. Environmental Protection Agency's (EPA) Green Power Partnership** provides data on the magnitude of purchases by its corporate, institutional, and government partners on an annual basis, which sheds light on non-residential purchaser trends. EPA works with more than 1,300 U.S. organizations to facilitate the purchase of green power. To be considered a "partner," the entity must buy green power in the voluntary market, meet certain thresholds for "new" renewables as a percent of the organization's annual electricity usage (3%-20%, depending on size), and execute a partnership agreement with EPA.
2. The **Center for Resource Solutions' (CRS) Green-e Energy certification program** provides data on the volume of renewable energy generation sold into voluntary markets and certified by the Green-e Energy program annually. All 2011 figures in this report are preliminary estimates. CRS maintains data on annual sales, by product

(e.g. green pricing) and customer type, number of customers, and location of supply and sales by state. The certification ensures that the renewable energy sold is from "new" resources, that sellers undergo marketing compliance review twice per year, that sales are verified annually, and that RECs are not double counted.[5] Not all voluntary RECs are Green-e Energy certified; though it was the largest certifier in the voluntary market at 23 million MWh of retail sales in 2010, or an estimated 65% of total voluntary market sales, and an estimated 99% of voluntary retail REC transactions.

3. The **U.S. Energy Information Administration** (EIA) collects data on green power sales and customers from utilities and marketers as part of its Form 861 (annual data) and Form 826 (monthly data). Not all competitive retailers report to EIA; therefore, these data underestimate sales, particularly in states with competitive retail markets. The monthly Form 826 collects from a sample of respondents compared to the annual Form 861 and data are released faster. EIA has published monthly Form 826 data beginning in January 2011 through December 2011. EIA's annual data for 2011 is expected to be released in mid-November 2012.
4. Regional **REC tracking systems**, initially created to ensure compliance with renewable portfolio standards (RPSs), are increasingly being used to verify voluntary REC market transactions. Public reports available from regional tracking systems provide the quantity of voluntary RECs issued to generators registered in their system, and in some cases provide the quantity of voluntary RECs retired.[6] Not all voluntary RECs transactions are verified in a tracking system and Green-e Energy's rules allow for a 21 month window to retire RECs for a calendar year.[7] Therefore, this data source presents an incomplete picture of total voluntary sales. Tracking systems operate primarily on a state or regional basis, often following the same boundaries as local regional transmission organizations (RTOs) or ISOs.

2. Voluntary Market Sales

In 2011, growth in voluntary market sales can be seen by examining the various available data sources, although none represent a comprehensive picture of the market. Both EPA and Green-e reported growth rates in renewable energy purchased or certified of about 20% in 2011. Data from REC tracking systems, which is the most comprehensive data source, but excludes transactions that occur outside of these systems, show about 40 million MWh of renewable energy in voluntary markets in 2011 (see Figure 1). While this figure is up 90% from 2010, much of the growth may be attributed to the increased use of REC tracking systems to verify voluntary market transactions, rather than an increase in the total market volume.

Data from tracking systems and other sources cannot be directly compared to market estimates prepared by NREL in the past. Historically, NREL has produced a comprehensive annual sales estimate, which has exceeded sales documented in tracking systems, EPA's Green Power Partnership, and CRS' Green-e Energy program. NREL's comprehensive estimate was drawn from gathering data directly from utility green pricing programs, competitive marketers, and REC brokers, supplemented with data from REC certifiers, ERCOT, and press releases describing large voluntary green power purchases. In 2010, tracking systems represented about 60% of the NREL estimated total market volume of about 35 million MWh.

EPA's Green Power Partnership saw continued growth in 2011, with sales increasing by 22% over 2010. Between 2006 and 2011, sales increased at a compound annual growth rate of 26%. Data through June 31, 2012 show that 2012 sales are up 10% relative to year-end 2011, indicating that if the second half of 2012 performs similarly, 2011-2012 growth should be similar to the year-to-year pattern in 2010-2011. In 2011, sales growth was driven by an increase in the average and median purchase size of partners with large load, which increased by 17% and 37%, respectively.[8]

In 2011, Green-e Energy retail sales increased to an estimated 27.9 million MWh, or 21%, similar to growth from 2009 to 2010. Green-e Energy retail sales increased to 23.1 million MWh in 2010, up from 18.7 million MWh in 2009, representing a 24% increase. From 2006 to 2011 Green-e Energy retail sales have seen a compound annual growth rate of 44%.

In 2011, estimated voluntary sales in tracking systems increased to 39.5 million MWh from 20.8 million MWh in 2010 (see Figure 2). From 2007 to 2011, estimated voluntary sales in tracking systems increased at a compound annual growth rate of 114%. However, this growth likely reflects the increased use of tracking systems as a verification tool for voluntary market RECs, not just market growth. Tracking system use is becoming more common because generators want to participate in the RPS compliance markets, which often require generators to use a tracking system, and generators and others are seeking to simplify the Green-e Energy verification and audit process.

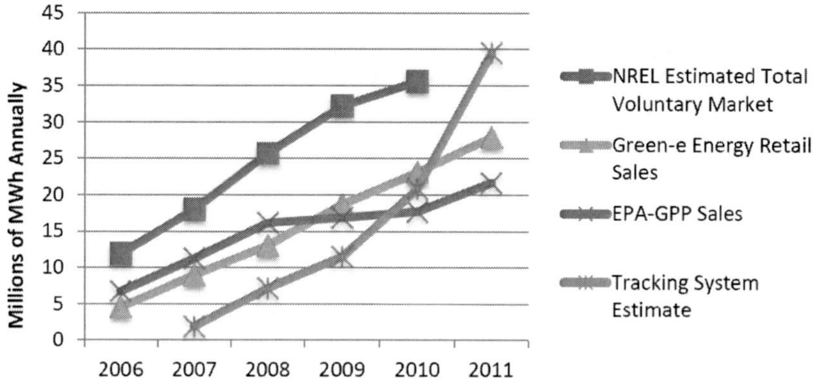

Sources: EIA 2012, EPA 2012a, Heeter and Bird 2011, Terada 2012.

Figure 1. Measurements of the voluntary market for renewable energy by recording entity, 2006-2011.

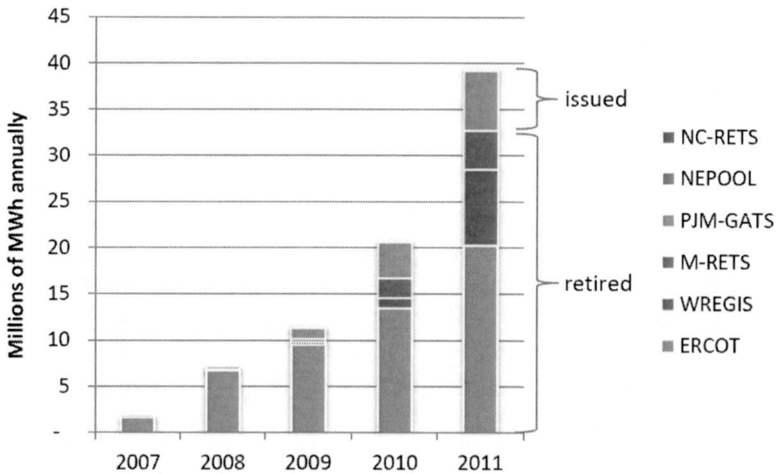

Note: The ERCOT, M-RETS, and WREGIS tracking systems report data for retired RECs; NC-RETS, NEPOOL, and PJM-GATS do not report retired REC data and are thus represented with issued REC data.

Figure 2. Voluntary REC market data from tracking systems, 2007-2011.

Estimates of market volume from tracking systems is based on data from the eight state or regional tracking systems in Table 1 (see also Figure 3). Data on retired certificates are used when available, but issued certificates are used as a proxy for systems that do not publicly report retirements. While all tracking systems provide the number of issued certificates publically, not all publically report the number of retired certificates (Table 1). Because voluntary certificates may be issued in one year but retired in a future year, the issued data from NC-RETS, NEPOOL, and PJM-GATS may overestimate retired voluntary certificates.

Table 1. Public Reporting by REC Tracking Systems

Tracking System	Number of Issued Certificates	Number of Retired Certificates
ERCOT	Yes	Yes
NEPOOL-GIS	Yes	No
PJM-GATS	Yes	No
WREGIS	Yes	Yes
M-RETS	Yes	Yes
NAR	Yes	No
MI-RECS	Yes	No
NC-RETS	Yes	No

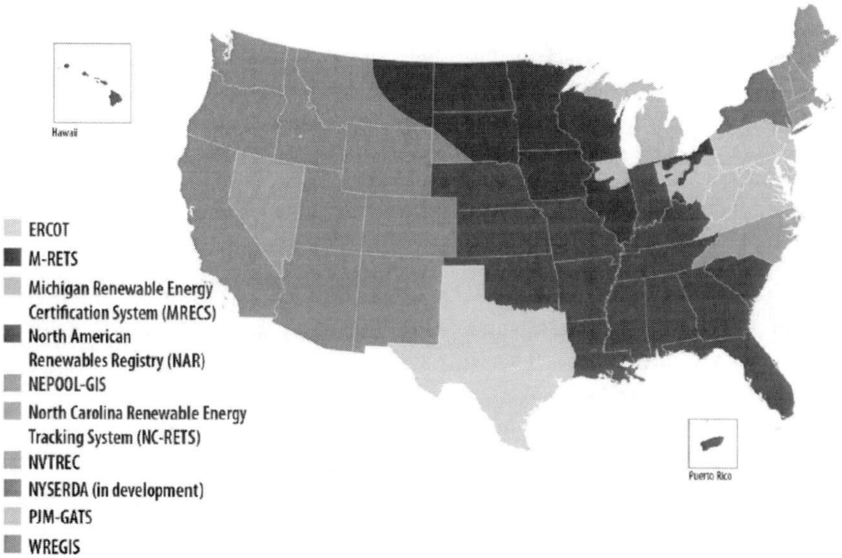

Source: Updated from ETNNA 2011.

Note: NAR covers states and provinces not covered by a NYSE Blue tracking system.

Figure 3. Renewable energy tracking systems in the United States.

WREGIS saw the largest increase in voluntary REC retirements, increasing from 1.1 million MWh in 2010 to 8.3 million MWh in 2011. An increasing number of utilities are using WREGIS to track their voluntary green pricing programs, and these programs are also growing on a MWh basis (Coon 2012). ERCOT saw about a 50% increase from 13.4 million MWh in 2010 to 20.2 million in 2011. ERCOT represented nearly half of the entire volume of transactions verified in REC tracking systems in 2011. The Texas market is further discussed in Section 2.2.

2.1. Voluntary Sales by Market Type

Historically, REC markets have represented the largest share of the voluntary market—larger than the volumes of green power sold through utility programs or competitive retail suppliers.[9] In 2010, 56% of all green power sales occurred as REC only transactions that are separate from electricity sales.

REC-only products appeal to nonresidential customers because of their flexibility and the potential cost savings created by sourcing from renewable energy projects in more favorable resource locations. Also, the electricity does not have to be delivered directly to the customer, which lowers transaction costs. For commercial and institutional customers that operate facilities in multiple locations across the country, RECs may also provide a more efficient green power sourcing solution than working with utilities in each individual utility territory. RECs may have a smaller appeal to residential customers because they may not be aware that RECs are available or may not understand what they convey.

EPA's Green Power Partnership tracks which sources partners are using to meet their voluntary commitments. In 2011, REC purchases represented 74% of total green power purchases under the Partnership, a level that has remained fairly steady since 2007. Sales through June 30, 2012 show that RECs represent 75% of total purchases in the Partnership (Figure 4).

The Green-e Energy certification program began with a focus on competitive electricity products, but has increasingly certified REC sales (Figure 5). In 2004, REC sales represented 4% of total Green-e Energy sales, while in 2011, REC sales represented approximately 84% of total Green-e Energy sales.

Utility green pricing sales for 2011 are available for those utilities that report to EIA under Form 826.[10] Figure 6 presents 2011 sales growth for a sample of green pricing programs based on EIA data. Note that while NREL has included "Top 10" rankings of programs in past reports, Figure 6 does not represent a similar ranking of top programs because data are incomplete.[11] Some green pricing programs, including Austin Energy, do not report to EIA under the 826. In the sample of programs examined from EIA 826 data, some programs increased sales while others saw sales decline, with the median sales growth at 1% from 2010 to 2011. The largest programs for which 2011 data is available saw increases in annual sales ranging from 2% to 9%: Portland General Electric increased by 740,000 MWh, PacifiCorp by 606,000 MWh, SMUD by 409,000 MWh, and Puget Sound Energy by 342,000 MWh.

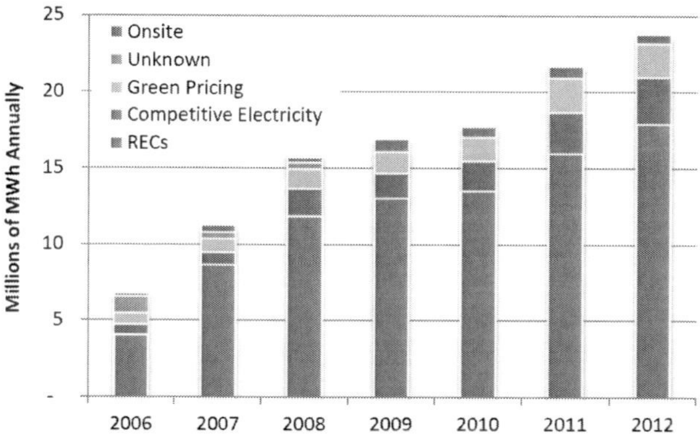

Source: EPA 2012a.

Note: 2012 data are through June 30, 2012.

Figure 4. EPA Green Power Partnership annual voluntary sales by classification system, 2006-2012.

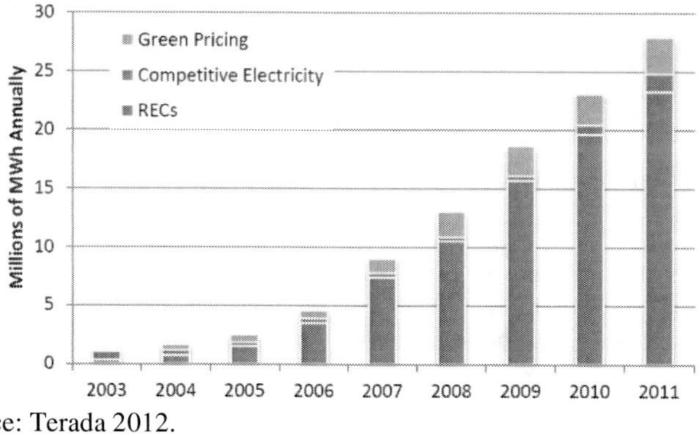

Source: Terada 2012.

Figure 5. Green-e Energy certified retail sales by product type, 2003-2011.

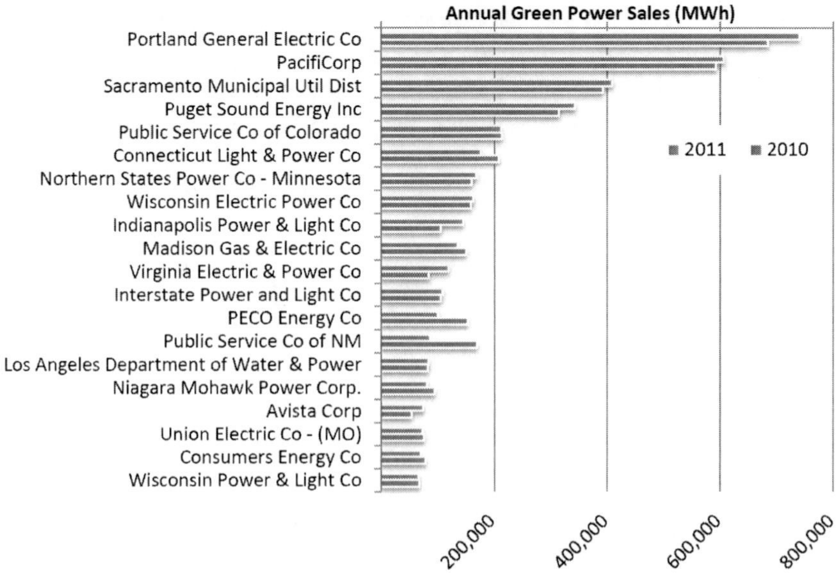

Sources: EIA 2012, EIA 2011.

Figure 6. Sample of annual green power sales by utility, 2010-2011.

2.2. Regional Focus

Utility green power and competitive market sales are predominant in certain states.[12] The top states in terms of total green power sales include California, Illinois, Maryland, Oregon, Texas, and Washington (Figure 7). State data on utility and competitive market sales for 2010 are publically available from the Energy Information Administration (EIA) (Figure 13 and Appendix B), which collects the data directly as part of its Form 861. However, because not all competitive retailers report to EIA, these data underestimate sales, particularly in states with competitive retail markets, such as Texas.[13]

Spotlight: Texas

In Texas, an increasing number of green power retail suppliers has in part driven growth in voluntary retirements of RECs. According to data reported by competitive marketers and utility green power programs to ERCOT, voluntary REC retirements in Texas increased by 29% between 2010 and 2011, from 11.8 million MWh in 2010 to 15.3 million MWh in 2010 (ERCOT 2012). In addition, some RECs issued in previous years were retired in 2011, raising the overall market total. In 2011, an additional 4.9 million MWh generated in 2009 and 2010 were also retired (Figure 8), for a total of 20.2 million MWh of voluntary retirements in 2011. In 2011, voluntary retirements in Texas surpassed compliance retirements for the fourth year in a row (ERCOT 2012).

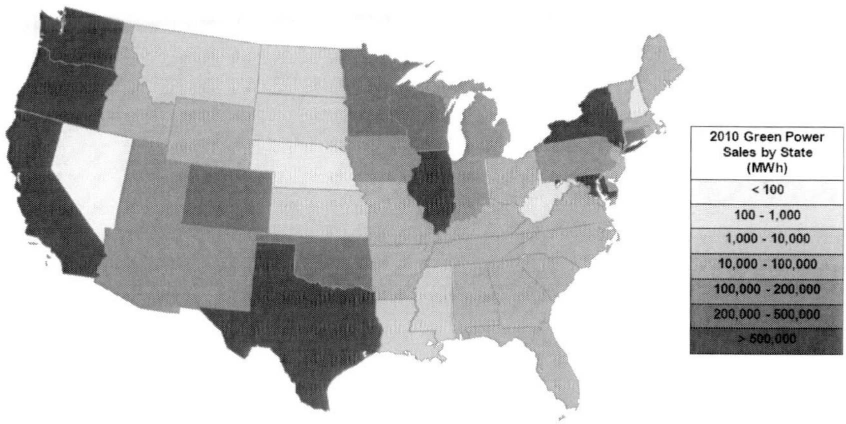

Source: EIA 2011.

Figure 7. Green power sales from EIA reporting entities, 2010.

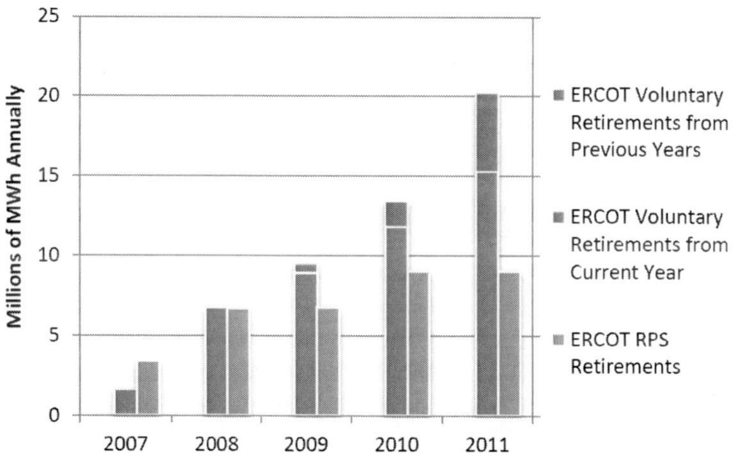

Sources: ERCOT 2012, ERCOT 2011, ERCOT 2010, ERCOT 2009, ERCOT 2008.

Figure 8. ERCOT voluntary and compliance retirements, 2007-2011.

Table 2. 20 Largest ERCOT Accounts by Voluntary REC Retirements, 2010

Account Holder	Number of RECs Voluntary Retired in 2010
Reliant Energy Retail, LLC	2,581,954
Green Mountain Energy Company	1,714,209
CPS Energy	1,605,437
NextEra Energy Power Marketing, LLC	978,693
Austin Energy	855,020
Renewable Choice Energy	741,968
First Choice Power Special Purpose, LP	395,529
SUEZ Energy Resources NA, Inc. - Super Green	329,037
SUEZ Energy Resources NA, Inc.	323,095
Washington Gas Energy Services	276,931
Nexant Clean Energy Solutions	268,528
Capricorn Ridge Wind II, LLC	231,726
Direct Energy LP - Green-e Retirements	224,694
Denton Municipal Electric	141,732
Gexa Corp.	139,389
Community Energy, Inc.	133,611
Clean Currents	122,327
3Degrees Group, Inc. (Green-e)	80,000
Green Mountain Energy Commercial Service	78,880
Dell USA, LP	64,095

Source: ERCOT 2012.

The Texas Public Utility Commission (PUC) has encouraged public access to REC market data by requiring ERCOT to report annually the aggregate quantity of RECs retired for voluntary and compliance purposes. In the current reporting year, confidentiality is ensured to account holders, which may be retiring compliance or voluntary RECs, but after one year, ERCOT documents how many RECs were retired by each account holder.[14] ERCOT data on the voluntary retirements by account holder for 2010 show that the top 5 largest voluntary retirements were completed by Reliant Energy, Green Mountain Energy, CPS Energy (San Antonio), NextEra, and Austin Energy. Table 2 shows that ERCOT account holders may be entities that operate and are selling RECs generated in Texas to end-consumers outside of Texas. From 2009 to 2010, the largest increases in retired voluntary RECs came from Reliant Energy, Green Mountain Energy, NextEra, Renewable Choice, and First Choice Power.

2.3. Residential and Nonresidential Customer Sales

No data are available for 2011 on sales by customer class, but some earlier data are available. Sales to nonresidential customers have represented between 73% and 76% of total sales from 2006 to 2010 (Figure 9). Figure 15 delineates green power sales by customer segment over time.

EIA's 2010 data show that while non-residential sales dominate in some states, residential sales contribute more than 50% of green power sales reported to EIA in Texas, Oregon, New York, California, Washington, and Wisconsin (Figure 10). Texas in particular has seen a rapid increase and uptake in the number of customers by wind energy from retail electricity suppliers.

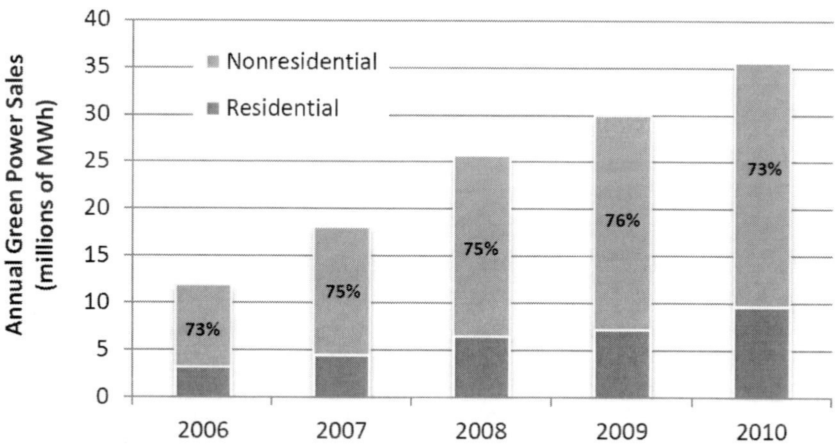

Source: Heeter and Bird 2011.

Figure 9. Residential and non-residential voluntary sales, 2006–2010.

3. Customer Participation

Comprehensive data on customer participation in 2011 are unavailable, but data are available for some utilities, as well as from EPA's Green Power Partnership. For the largest 20 programs (in terms of customers) that report in the EIA 826 Form, customer participation totaled roughly 488,000 in 2011, up 6.5% from 458,000 in 2010 (EIA 861 data) (Figure 11). The largest four programs (in terms of green power customers) that reported data in the EIA

826, Portland General Electric, PacifiCorp, SMUD, and Puget Sound Energy, saw customer gains in 2011 ranging from 5% to 10%.

The average participation rate for the largest 20 programs in terms of green power customers in 2011 was 2.9%, ranging from 1.9% to 9.8% (Figure 12) while the average participation rate for all programs reported in the EIA 826 Form in 2011 was 0.9%.

The number of participants in EPA's Green Power Partnership began to rise in the first half of 2012, after seeing nearly flat net participation from 2010 to 2011 (Figure 13). Notably, in 2011, a larger number (142) of small partners left the program than in previous years; although net participation was nearly flat, new organizations did join the program (EPA 2012a). From 2011 through June 2012, net participation grew by 8% to 1,407 partners.

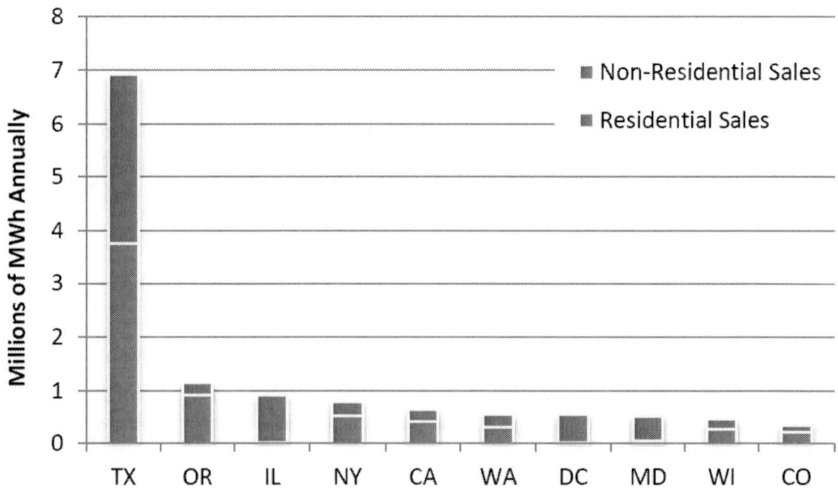

Source: EIA 2011.

Figure 10. Electricity sales by state for green pricing programs, 2010.

Two new large purchasers joined the EPA Green Power Partnership in 2012: Microsoft and McDonald's USA. Microsoft is purchasing 1,120 thousand MWh of green power annually (ranked third) and McDonalds is purchasing 306 thousand MWh of green power annually (ranked thirteenth), as of July 2012. The top purchasers in EPA's Green Power Partnership are listed in Appendix A. NREL has estimated that 1.8 million electricity customers nationwide purchased green power products in 2010 through regulated utility companies, from green power marketers in a competitive-market setting, or in the form of RECs (Table 3).[15]

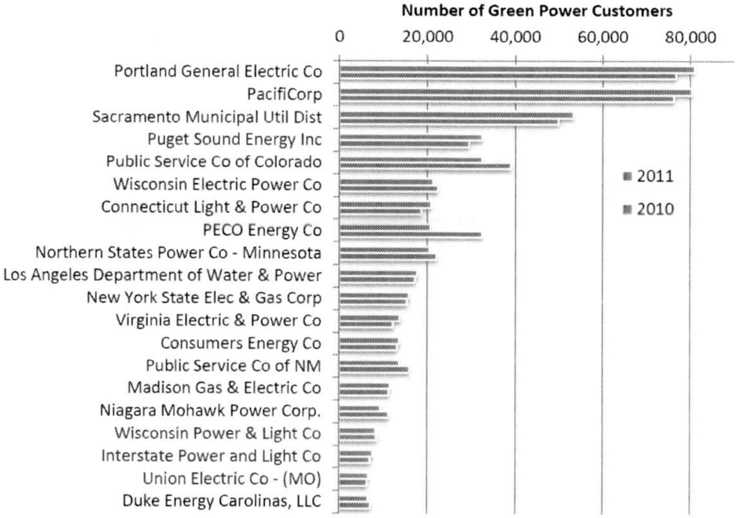

Sources: EIA 2012, EIA 2011.

Figure 11. Sample of annual green power customers by utility, 2010-2011.

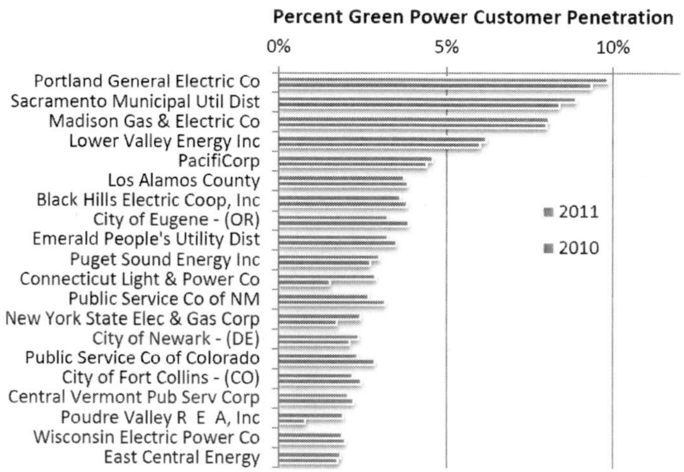

Sources: EIA 2012, EIA 2011.

Figure 12. Sample of percent green power customer penetration by utility, 2010-2011.

This represented a 25% increase in participation from 2009. While not as strong as 2009 growth of 44%, the 2010 growth was stronger than previous years (2006-2008). Participation in competitive markets increased about 45% primarily due to substantial customer increases of one marketer operating in states with retail competition. REC market participation also increased considerably, due to an increase in residential customer participation. Participation in utility green pricing programs was up slightly in 2010, due to increases in residential customer participation. However, no new figures are available for 2011.

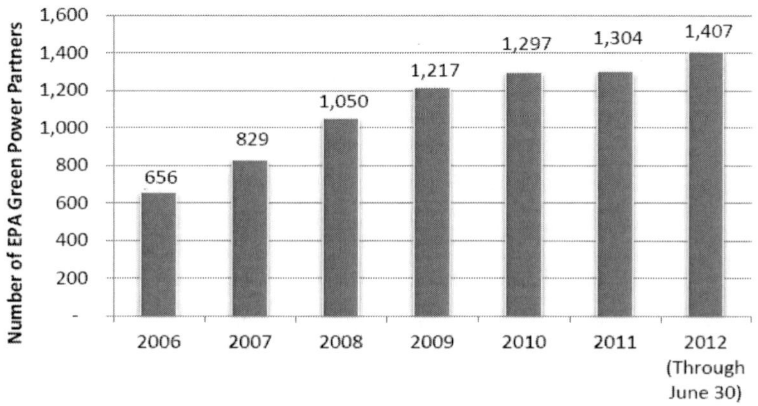

Source: EPA 2012a.

Figure 13. Number of participants in EPA's Green Power Partnership.

Table 3. Estimated Cumulative Green Power Customers by Market Segment, 2006–2010

	2006	2007	2008	2009	2010
Utility Green Pricing	**490,000**	**550,000**	**550,000**	**550,000**	**570,000**
Residential	*470,800*	*526,700*	*519,700*	*526,300*	*544,700*
Nonresidential	*15,500*	*20,200*	*26,100*	*26,000*	*22,900*
% Residential Growth	*23%*	*12%*	*-1%*	*1%*	*4%*
% Nonresidential Growth	*37%*	*30%*	*29%*	*(1%)*	*(12%)*
Competitive Market	**~ 210,000**	**300,000**	**390,000**	**830,000**	**~ 1,200,000**
Voluntary REC Market	**~ 10,000**	**> 10,000**	**30,000**	**< 20,000**	**> 60,000**
Retail Total	**~ 710,000**	**~ 860,000**	**~ 970,000**	**~ 1,400,000**	**~ 1,830,000**
% Change	*~ 22%*	*~ 21%*	*~ 13%*	*~ 44%*	*~ 25%*

Note: In some cases, estimates have been revised from those reported in previous NREL reports as updated data have become available. Totals may not add due to rounding.

4. Community Solar Programs

Increasingly, utilities and third parties are developing community solar programs[16] that allow customers to purchase a share of a renewable system developed in the local community. These types of programs represent the vast majority of new green pricing programs.

Under community solar programs,[17] customers receive the benefits of the energy that is produced by their share. For example, the Holy Cross Energy solar project in El Jebel, Colorado, is an 80 kW photovoltaic system supported by 18 community participants who purchase shares at an upfront cost of $3.15/W ($3,150/kW) and then receive a credit on their bill each month at a rate of $0.11/kWh (Green Power Network 2010). Typically, community solar programs require an upfront investment in a "share" or "panel" of the project, which can cost hundreds of dollars. However, that is not always the case. Delta-Montrose Electric Association's Community Solar Array program sells shares in $10 increments (Green Power Network 2011).

Advantages of community solar programs include potential cost savings due to economies of scale, as community solar programs may be able to offer a lower price per watt. Community solar programs typically also allow consumers to keep their shares if they move within the utility's service territory. Also, community solar may be an option for consumers who lack sufficient roof space or exposure to warrant installation of a system on their homes. For more on these issues, see Heeter and McLaren (forthcoming).

The majority of new utility programs introduced in recent years have been community solar programs. In 2011, 10 programs were introduced, and as of August 2012, an additional 8 programs had begun in 2012 (see Appendix C). U.S. community solar programs have a combined capacity of nearly 10,400 kW as of August 2012, and that capacity has been growing (Figure 14). In 2011, the Salt River Project (Arizona) began a 2,000 kW program for residential participants.[18] Two additional large (>1,000 kW) programs were developed in Arizona: Tuscon Electric Power's 1,600 kW program in 2011 and UniSource Energy Service's 1,720 kW program in 2012. San Miguel Power Association, working with the Clean Energy Collective (Colorado) also developed a large system (1,000 kW) in 2012.

In some cases, community solar programs have been enabled by state legislation. In May 2009, Washington passed SB 6170, which enables community solar participants to qualify for the state's production incentive program (DSIRE 2012). Projects up to 75 kW are eligible. The production incentive can range between $0.12/kWh and $0.54/kWh, depending on whether the project qualifies for certain local content multipliers, and each participant in a community solar project is eligible to receive the incentive, which is capped at $5,000 per year per participant (DSIRE 2012).

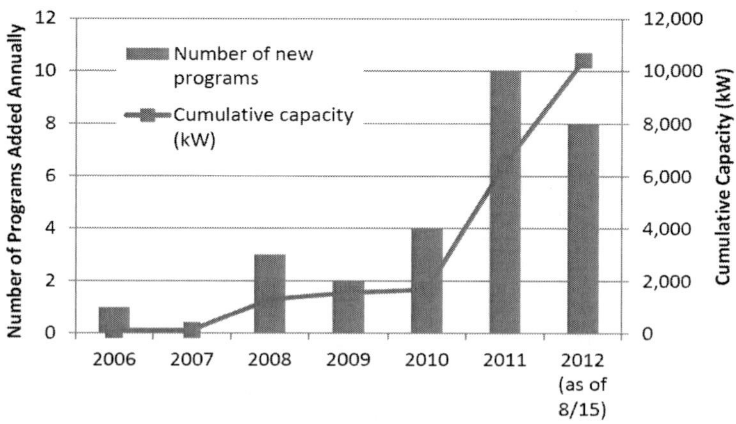

Figure 14. Number and cumulative capacity of community solar programs, January 2006–August 2012.

In Colorado, legislation was passed in June 2010 requiring that IOUs develop plans to acquire RECs from community solar gardens. The bill required the PUC to set a minimum and maximum purchase of electrical output for each utility. In order to comply, the state's largest utility, Xcel Energy, will offer a new Solar*Rewards Community incentive program. In June 2012, the Colorado PUC finalized rules pertaining to the incentive program. Under this new program Xcel will accept 4.5 MW of community solar electricity in 2012; it will pay $0.14 per kWh for small programs (10-50 kW), and $0.11 per kWh for medium programs (50-500 kW), then scaling down the payments over time after 3 MW of capacity has been installed. Community solar developers were invited to submit applications on August 15, 2012, and Xcel closed the application process after 30 minutes, as three times the 2012 capacity allotment had been submitted.

5. REC PRICING

In 2011, prices for voluntary RECs (both national any technology, national wind, and western wind) declined. Compliance REC pricing increased in a few states (primarily in the Northeast).

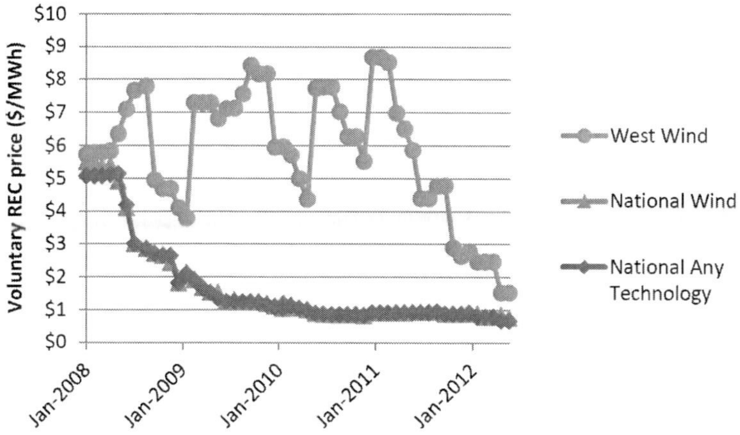

Sources: Spectron Group 2012.

Note: Plotted values are the last trade (if available) or the mid-point of bid and offer prices for the current or nearest compliance year for various state compliance RECs.

Figure 15. Voluntary REC prices, January 2008–June 2012.

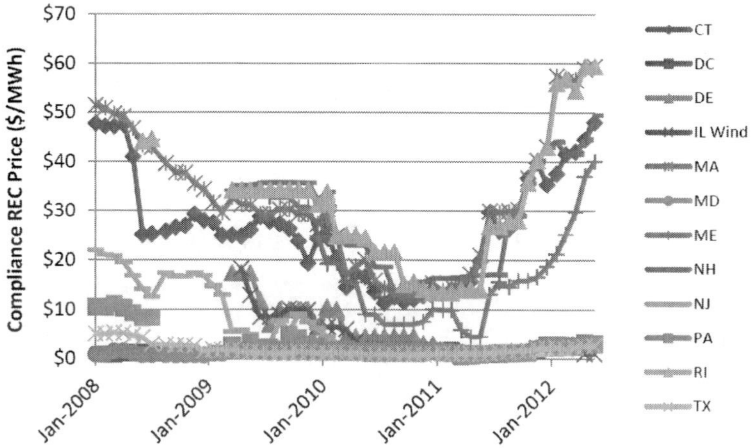

Sources: Spectron Group 2012.

Note: Plotted values are the last trade (if available) or the mid-point of bid and offer prices for the current or nearest compliance year for various state compliance RECs.

Figure 16. Compliance market (Tier 1) REC prices, January 2008–June 2012.

As shown in Figure 15, wholesale RECs used in voluntary markets have generally traded in the range of $1/MWh to $10/MWh (0.1¢/kWh to 1.0¢/kWh), based on available indicative broker data. In 2011 and early 2012, prices paid for nationally sourced voluntary RECs from any technology dropped to less than $1/MWh. Nationally sourced voluntary wind REC prices were comparable to nationally sourced voluntary RECs for any technology. Wind from the western United States earned higher prices, though West Wind fell to less than $2/MWh in 2012. The drop in West Wind pricing may be a result of utilities in California being on track to meet the state's RPS. Bloomberg New Energy Finance has noted that California's IOUs do not have a need for additional renewable procurement through the first compliance period (2011-2013) (BNEF 2012).

Compliance REC prices differ because of regional differences in renewable energy resource quality (e.g., wind speed) and the supply and demand of RECs. In 2011, compliance REC prices began to increase in some markets, particularly in the Northeast (Figure 5). Prices for Massachusetts, Rhode Island, New Hampshire, and Connecticut RECs increased in 2011 and the beginning of 2012 to between $50/MWh and $60/MWh. Other state markets have seen little to no change in REC pricing in recent years. REC prices remain under $5/MWh in Washington, D.C., Delaware, Illinois, Maryland, New Jersey, Ohio, Pennsylvania, and Texas. Figure 16 shows the wide variation in compliance REC prices among states for which data are available.

Conclusion and Observations

Voluntary REC markets continue to exhibit growth and provide a stimulus for renewable energy development. Green power markets provide an additional revenue stream for renewable energy projects and raise consumer awareness of the benefits of renewable energy. Based on this review, NREL has identified the following market trends:

- In 2011, several available data sources show that voluntary sales of renewable energy continued to increase. Both EPA's Green Power Partnership and CRS' Green-e Energy program experienced growth rates of around 20% The EPA's Green Power Partnership grew to 21.7 million MWh in 2011, a 22% increase over 2010. The CRS'

Green-e Energy program grew to an estimated 27.9 million MWh in 2011 for retail sales, an increase of 21% over 2010.

- Data from REC tracking systems, which is the most comprehensive data source, but excludes transactions that occur outside of these systems, show about 40 million MWh of renewable energy in voluntary markets in 2011, up from 20.8 million MWh in 2010, a growth rate of 90%. In 2010, tracking systems represented about 60% of the NREL estimated total market volume of about 35 million MWh. Growth in this sector is likely partially a result of increased use of tracking systems for voluntary market transactions. ERCOT and WREGIS saw the largest growth on a MWh basis in 2011; and ERCOT represents roughly half of all reported volumes.
- Based on data from EIA on a sample of utility programs, the largest 20 programs that report in the EIA 826 recorded a mix of green power sales increases and declines from 2010 to 2011. A few programs saw sales growth greater than 35%, while others saw losses of similar proportion; the median growth for the sample of programs was 1%.
- The majority of new utility programs introduced in recent years have been community solar programs that enable utility customers to purchase a share of a solar system and receive the benefits of the energy produced by their share. Since 2011, 18 new programs were introduced. Today, nearly 30 community solar programs have a combined capacity of more than 10 MW.
- Voluntary REC pricing fell to less than $1/MWh in 2011. Pricing has remained under $1/MWh through August 2012.

APPENDIX A. LEADING PURCHASERS IN THE EPA GREEN POWER PARTNERSHIP

Table A-1. Top 25 Purchasers in the EPA Green Power Partnership Program, as of July 5, 2012

Company & Rank	Annual Green Power Usage (kWh)	GP % of Total Electricity Use - 2012	Green Power Resources
1. Intel Corporation	2,798,660,169	89%	Biogas, Biomass, Small-hydro, Solar, Wind
2. Kohl's Department Stores	1,528,378,000	101%	Solar, Wind

Table A-1. (Continued)

Company & Rank	Annual Green Power Usage (kWh)	GP % of Total Electricity Use - 2012	Green Power Resources
3. Microsoft Corporation	1,120,000,000	46%	Biomass, Small-hydro, Wind
4. Wal-Mart Stores, Inc. / California and Texas Facilities	872,382,088	28%	Biogas, Solar, Wind
5. Whole Foods Market	800,257,623	107%	Solar, Wind
6. Lockheed Martin Corporation	546,399,457	30%	Biogas, Small-hydro, Solar, Wind
7. Staples	516,713,408	78%	Biogas, Solar, Wind
8. City of Houston, TX	438,000,000	35%	Wind
9. Starbucks	414,560,000	46%	Biogas, Geothermal, Small-hydro, Wind
10. City of Austin, TX	370,464,066	100%	Wind
11. BD	317,276,000	60%	Solar, Wind
12. Hilton Worldwide	315,000,000	94%	Small-hydro, Wind
13. McDonald's USA LLC	306,000,000	30%	Wind
14. U.S. Department of Energy	295,974,615	6%	Geothermal, Solar, Wind
15. City of Dallas, TX	295,883,744	40%	Wind
16. Cisco Systems, Inc.	268,644,637	27%	Wind
17. District of Columbia	267,042,489	50%	Wind
18. U.S. Air Force	265,777,826	3%	Biogas, Biomass, Solar, Wind
19. TD Bank, N.A.	261,827,000	100%	Wind
20. U.S. Environmental Protection Agency	261,056,565	101%	Biogas, Biomass, Solar, Wind

Source: EPA 2012b.

APPENDIX B. EIA ESTIMATED U.S. GREEN PRICING SALES BY STATE

Table B-1. EIA Estimated U.S. Green Pricing Sales by State (MWh), 2010

State	2009 Green Pricing Sales (MWh)	2010 Green Pricing Sales (MWh)
AK	-	-
AL	7,659	12,915
AZ	104,548	130,494
AR	18,497	21,797
CA	809,262	640,910
CO	345,377	344,256

State	2009 Green Pricing Sales (MWh)	2010 Green Pricing Sales (MWh)
CT	192,971	235,413
DC	29,612	554,524
DE	90,160	178,343
FL	14,983	13,716
GA	56,306	60,536
HI	-	-
ID	48,820	44,711
IL	25,181	916,181
IN	112,885	140,329
IA	34,458	111,365
KS	73,435	1,020
KY	42,685	19,734
LA	3,350	1,948
ME	17,862	15,787
MD	171,138	525,206
MA	83,746	93,863
MI	172,649	174,820
MN	276,516	238,770
MS	1,113	5,315
MO	58,890	77,542
MT	6,308	5,484
NE	15,067	209
NV	81	69
NH	-	308
NJ	18,369	29,312
NM	219,210	190,673
NY	358,271	785,378
NC	12,898	24,424
ND	34,761	7,387
OH	24,468	18,033
OK	231,508	302,669
OR	1,130,908	1,145,227
PA	300,515	196,510
RI	33,150	32,998
SC	28,351	25,836
SD	263	1,925
TN	73,160	63,870
TX	5,102,146	6,920,311
UT	180,173	186,522
VT	16,674	20,666
VA	23,584	85,313
WA	579,015	555,440
WV	855	177
WI	449,843	136,060
WY	42,935	37,537

Source: EIA 861.

Appendix C. Community Solar Offers

Table C-1. Historical Development of Community Solar Offers

Utility/Provider	Program Name	Program Size (kW)	Program Start Year
Ellensburg (WA)	Community Solar Project	111	2006
Sacramento Municipal Utility District (CA)	SolarShares	1,000	2008
Florida Keys Electric Cooperative (FL)	Simple Solar	117	2008
Ashland (OR)	Solar Pioneers II	64	2008
St. George (UT)	SunSmart	250	2009
Bainbridge Island (WA)	Solar for Sakai	5	2009
Holy Cross Energy/Clean Energy Collective (CO)	Mid Valley Solar Array (El Jebel)	80	2010
University Park Community Solar LLC (MD)	University Park Solar	23	2010
Corvallis (OR)	Corvallis OR, SunSlice Deal	2	2010
Okanogan County Elec. Coop. (WA)	OCEC Community Solar	20.2	2010
Salt River Project (AZ)	SRP EarthWise & Copper Crossing Solar Ranch	2,000	2011
Trico Electric (AZ)	Trico SunWatts Sun Farm	193	2011
Delta-Montrose Electric Association (CO)	Community Solar Array	20	2011
Holy Cross Energy/Clean Energy Collective (CO)	Garfield County Array	858	2011
Berea Utilities (KY)	Berea Solar Farm	14	2011
Edmonds (WA)	Edmonds Community Solar Cooperative	4	2011
Okanogan County Elec. Coop. (WA)	Winthrop Community Solar	23	2011
Poulsbo Project (WA)	Poulsbo Middle School	75	2011
Seattle City Light (WA)	Community Solar	24	2011
Tuscon Electric Power (AZ)	Bright Tucson Community Solar Program	1,600	2011
UniSource Energy Services (AZ)	Bright Arizona Community Solar	1,720	2012
Colorado Springs (CO)	Community Solar Garden Facility Incentive Program	500	2012
Poudre Valley REA/Clean Energy Collective (CO)	Community Solar	115	2012
Brewster Community Solar Garden Coop. Inc. (MA)	Brewster Community Solar Garden	345.6	2012
Olympia (WA)	Olympia WA, SunSlice Deal	75	2012

Utility/Provider	Program Name	Program Size (kW)	Program Start Year
Acorn Energy Coop. (VT)	Acorn Energy Solar One	150	2012
United Power (CO)	Sol Partners Coop. Solar Farm	10	2012
San Miguel Power Association/Clean Energy Collective (CO)	SMPA Community Solar	1,000	2012

REFERENCES

Bird, L., Kreycik, C. & Friedman, B. (September 2009). *Green Power Marketing in the United States: A Status Report (2008 Data)*. NREL/TP-6A2-46581. Golden, CO: National Renewable Energy Laboratory, September. http://www.nrel.gov/docs/fy09osti/46581.pdf. Accessed October 4, 2011.

Bird, L. & Sumner, J. (September 2010). *Green Power Marketing in the United States: A Status Report (2009 Data)*. NREL/TP-6A20-4940. Golden, CO: National Renewable Energy Laboratory. http://www.nrel.gov/docs/fy11osti/49403.pdf. Accessed September 2011.

Bloomberg New Energy Finance (BNEF). (2012). U.S. Renewable Energy – Research Note: California renewable compliance: up to the brim. 22 June.

Clean Energy Collective (CEC). (2012). "*Bringing down the cost of solar through community shared solar*." Webinar presentation. http://www.icleiusa.org/action-center/learn-from-others/bringing-down-the-cost-of-solar-through-community-shared-solar-programs. Accessed August 14, 2012.

Coon, A. (17 August 2012). E-mail. *Western Electricity Coordinating Council*. Salt Lake City, Utah.

Database of State Incentives for Renewables & Efficiency (DSIRE). (2012). www.dsireusa.org. Accessed August 17, 2012.

Department of Energy (DOE). (2012). "*Guide to Community Shared Solar: Utility, Private, and Non-profit Project Development*. http://www1.eere.energy. Accessed August 14, 2012.

Electricity Information Agency (EIA). (2012). "*Monthly Electric Utility Sales and Revenue Data*." http://www.eia.gov/cneaf/electricity. Accessed September 4, 2012.

EIA. (2011). "Annual Electric Utility Data File – EIA-861." http://www.eia.gov/electricity. Accessed September 4, 2012.

Electric Reliability Council of Texas, Inc. (ERCOT). (2012). *ERCOT's 2011 Annual Report on the Texas Renewable Energy Credit Trading Program.* Project No. 27706. Texas Public Utilities Commission. https://www.texasrenewables.com/reports.asp Accessed September 4, 2012.

ERCOT. (2011). *ERCOT's 2010 Annual Report on the Texas Renewable Energy Credit Trading Program.* Project No. 27706. Texas Public Utilities Commission. https://www.texasrenewables.com/reports.asp. Accessed September 4, 2012.

ERCOT. (2010). *ERCOT's 2009 Annual Report on the Texas Renewable Energy Credit Trading Program.* Project No. 27706. Texas Public Utilities Commission. https://www.texasrenewables.com/reports.asp. Accessed September 4, 2012.

ERCOT. (2009). *ERCOT's 2008 Annual Report on the Texas Renewable Energy Credit Trading Program.* Project No. 27706. Texas Public Utilities Commission. https://www.texasrenewables.com/reports.asp. Accessed September 4, 2012.

ERCOT. (2008). *ERCOT's 2007 Annual Report on the Texas Renewable Energy Credit Trading Program.* Project No. 27706. Texas Public Utilities Commission. https://www.texasrenewables.com/reports.asp. Accessed September 4, 2012.

Environmental Protection Agency (EPA). (2012a). E-mail. Data on Green Power Partnership.

EPA. (2012b). Green Power Partnership – National Top 50. http://www.epa.gov/greenpower/toplists/top50.htm. Accessed April 2011.

Green Power Network. (2010). "Holy Cross Energy Launches 80 kW Community Solar Program." http://apps3.eere.energy. Accessed October 3, 2011.

Green Power Network. (2011). "Delta-Montrose Electric Association (Colo.) Launches Community Solar." http://apps3.eere.energy. Accessed October 3, 2011.

Heeter, J. & Bird, L. (October 2011). Status and Trends in U.S. Compliance and Voluntary Renewable Energy Certificate Markets (2010 Data). NREL/TP-6A20-52925. Golden, Colorado: National Renewable Energy Laboratory. http://apps3.eere.energy. Accessed August 17, 2012.

Heeter, J. & McLaren, J. (forthcoming). Innovations in Voluntary Renewable Energy Procurement: Methods for Expanding Access and Lowering Cost for Communities, Governments, and Businesses. Golden, Colorado: National Renewable Energy Laboratory.

Spectron Group. (2012). Data accessed through Vantage database. London, UK. www.spectrongroup.com. Accessed August 15, 2012.

Terada, R. (14 September 2012). Email. Center for Resource Solutions, San Francisco, CA.

End Notes

[1] A REC is issued to 1 MWh of renewable generation, but is retired when it has been used to meet RPS compliance or to make a voluntary claim.

[2] States with competitive offerings include Connecticut, Delaware, Illinois, Maine, Maryland, Massachusetts, Michigan, New Jersey, New York, Ohio, Pennsylvania, Rhode Island, Texas, and Washington, D.C.

[3] For a list of companies offering voluntary REC products, see the DOE's Green Power Network website: http://apps3.eere.energy.

[4] Voluntary market data from previous years is captured in earlier versions of the report including: Heeter and Bird 2011; Bird and Sumner 2010; and Bird et al. 2009.

[5] The Green-e Energy National Standard and EPA's Green Power Partnership both consider "new" projects to be those that have online within 15 years prior to the sale of the green power.

[6] A REC is *issued* to 1 MWh of renewable generation, but is *retired* when it has been used to meet RPS compliance or to make a voluntary claim.

[7] For example, for certified sales made in calendar year 2011, generation could have come from July 2010 through March 2012.

[8] Large partners are defined as partners with more than 100,000 MWh of load annually.

[9] The REC market sales figures reflect sales to end-use customers separate from electricity. RECs bundled with electricity and sold to end-use customers through utility green pricing programs or in competitive electricity markets are counted in other categories.

[10] EIA data were spot-checked for reporting errors but otherwise not validated by NREL. Madison Gas & Electric Company provided a revised 2011 green power sales figure to NREL.

[11] Previous NREL lists have also presented data on a program basis rather than a utility basis (e.g. sales for a utility program that operates in multiple states have been aggregated).

[12] Data on the geographic source of unbundled RECs is not available from EIA.

[13] According to EIA, Form 861 is completed by "electric utilities, wholesale power marketers (registered with the Federal Energy Regulatory Commission), energy service providers (registered with the States), and electric power producers. Responses are collected at the business level (not at the holding company level) (EIA 2011)."

[14] ERCOT's Annual Report on the Texas Renewable Energy Credit Trading Program can be found at https://www.texasrenewables.com/reports.asp.

[15] There is greater uncertainty in NREL customer estimates for competitive and REC markets because of data limitations. Generally, NREL estimates are consistent with the EIA estimates when adjusted for customers in Ohio who participated in community aggregations in 2005 and earlier. NREL excluded these customers from our estimates because they purchase products with very low renewable energy content (1%–2%).

[16] For an examination of how to develop a community solar project, see DOE 2012.

[17] While the term "community solar program" encompasses other business models (such as bulk purchasing of photovoltaic equipment), this report addresses only programs where participants purchase a share of the solar project and receive credit for their solar production.

[18] Salt River Projects solar array includes an additional 7,840 kW reserved for local schools.

In: Voluntary Markets in Renewable Energy ISBN: 978-1-62417-561-9
Editor: Jamir S. Fuller

Chapter 2

INNOVATIONS IN VOLUNTARY RENEWABLE ENERGY PROCUREMENT: METHODS FOR EXPANDING ACCESS AND LOWERING COST FOR COMMUNITIES, GOVERNMENTS, AND BUSINESSES*

Jenny Heeter and Joyce McLaren

1. INTRODUCTION

This guide explores five innovative options for voluntarily procuring renewable energy generation or systems. The options and their descriptions are shown in Table 1. These methods can be replicated by a variety of stakeholders—including local governments, not-for-profit organizations, businesses, and utilities. The three decision trees in Figure 1 indicate which method may best suit a particular stakeholder's needs.

The options described in this guide offer a variety of benefits, including lower cost, simplified transactions, expanded access to green energy, and increased impact (see Text Box 1). Each chapter describes one option, provides examples of its use, lays out design and implementation considerations, and discusses benefits and challenges.

---

* This is an edited, reformatted and augmented version of the National Renewable Energy Laboratory, NREL/TP-6A20-54991, dated September 2012.

New purchasing methods have been created by utilities, local governments, businesses, and others to expand access, lower the cost, and simplify the process of procuring voluntary renewables.

Table 1. Five Methods to Encourage Voluntary Procurement of Renewable Energy Generation or Systems

Method	Description	Examples
Community Choice Aggregation (CCA) *Section 2 (page 9)*	CCA programs allow communities to collectively choose the source of their electricity generation while maintaining transmission and distribution service from the existing provider. CCA programs may or may not select renewable energy supply; this paper focuses on such programs that do provide renewable energy.	Marin Clean Energy (CA) Oak Park Community Choice Aggregation (IL)
Community wind and solar programs *Section 3 (page 14)*	Community programs allow participants to invest in renewable energy projects; these programs include community solar and wind programs. Programs typically fund new local solar or wind capacity; each participant purchases a portion of the capacity.	Community solar: Sacramento Municipal UtilityDistrict's SolarShares Program (CA) Community wind: Huerfano River Wind Project (CO)
Green power challenges and local collaborative electricity procurement *Section 4 (page 18)*	Green power challenges are media campaigns sponsored by local governments, utilities, third-party marketers, and/or environmental organizations to encourage increased levels of voluntary green power purchasing in a community. Local collaborative electricity procurement aggregates the demand of a few organizations in deregulated electricity markets to procure renewable electricity.	Green power challenge: Lake Oswego (OR) Local collaborative electricity procurement: Groundswell's Community Power Project (D.C.)
Bulk purchasing of on-site systems *Section 5 (page 21)*	Bulk purchasing programs aggregate demand from a group of individuals or companies interested in installing on-site systems, attracting reduced prices from vendors, and simplifying the purchasing process.	Solarize Portland (OR) One Block Off the Grid (National)
Reverse auctions for renewable energy *Section 6 (page 25)*	Reverse auctions are mechanisms for one or more buyers to request a renewable energy product and have many sellers bid to provide the product; the bidder that can provide the product for the lowest price wins the auction.	World Energy Solutions (National)

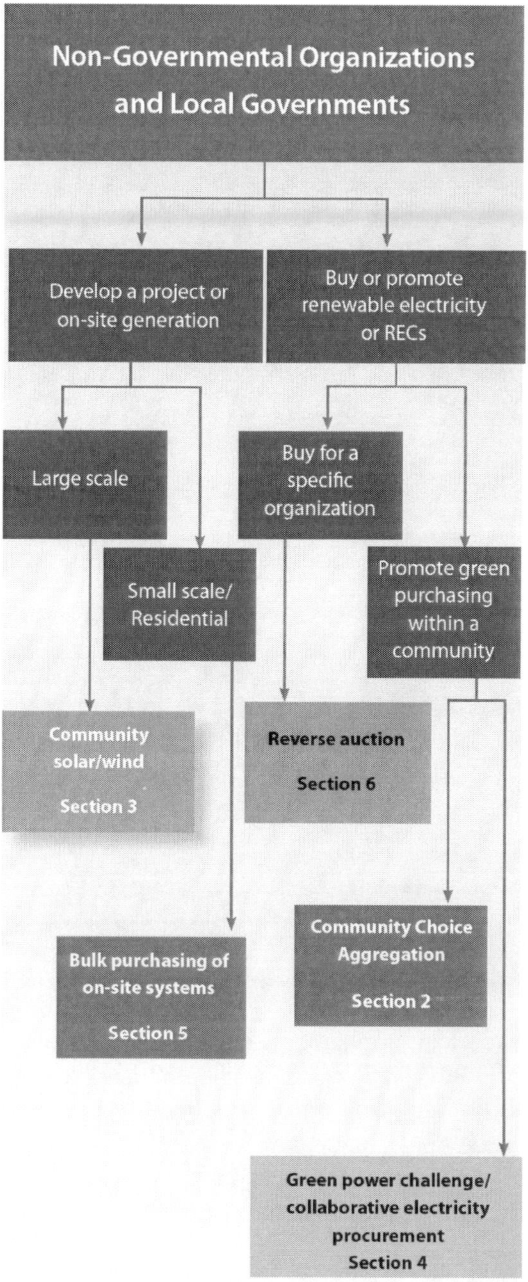

Figure 1. Decision trees for selecting an innovative purchasing method.

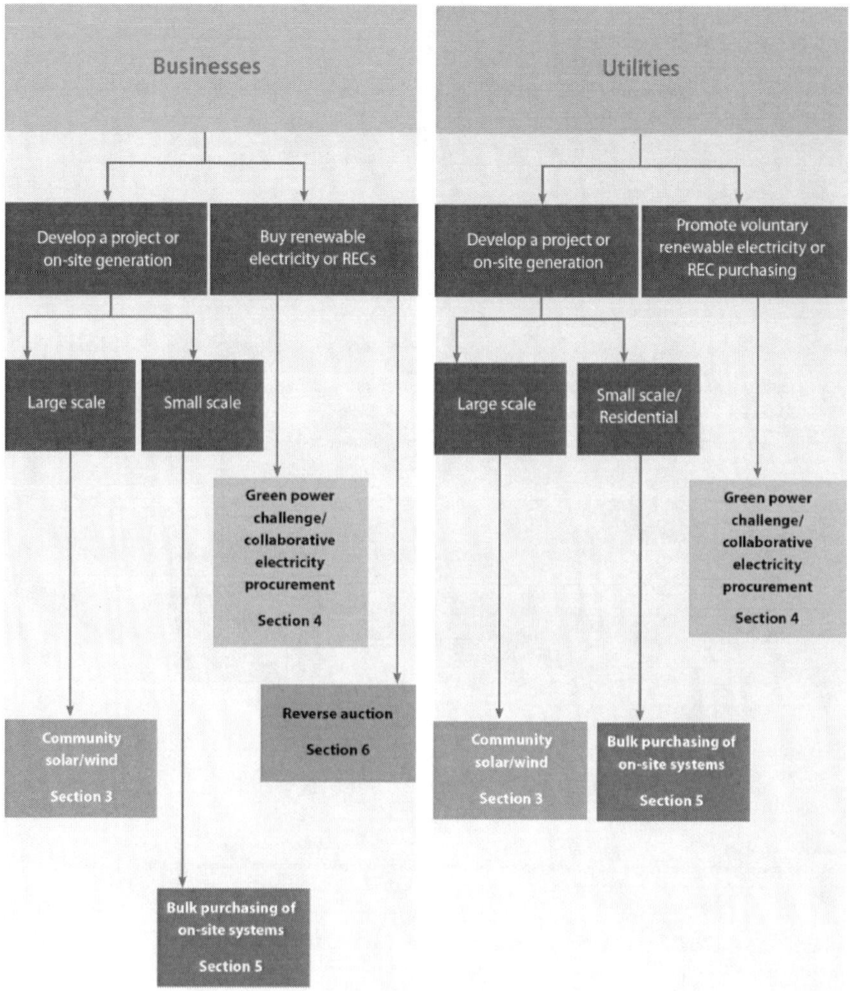

Voluntary purchasing of renewable energy—through utility green pricing programs, competitive suppliers, unbundled renewable energy certificate (REC) markets, and installation of on-site systems—is on the rise. Individual, corporate, and institutional purchasing of renewable energy has expanded rapidly in recent years, increasing from 11.9 million MWh in 2006 to 35.6 million MWh in 2010 (Heeter and Bird 2011). On-site solar photovoltaic installations have also been increasing: In 2011, the capacity of distributed grid-connected PV doubled compared to 2010, reaching 1.3 GWdc (Sherwood 2012).

Text Box 1. Benefits of Innovative Mechanisms

Decrease costs of obtaining renewable energy. Bulk purchasing of on-site systems and reverse auctions in particular are models built on the assumption of providing lower cost, but Community Choice Aggregation (CCA) programs and community solar programs may also provide renewable energy at lower cost.

Expand access to renewable energy. Innovative mechanisms work to increase access to renewable energy by making it easy to participate or by addressing common barriers. CCA programs with opt-out[1] structures ensure that a portion of a customer's electricity comes from renewable energy, unless that customer opts out. Community solar programs provide options to renters and to homeowners with shaded roofs.

Facilitate the process of buying renewables. Bulk purchasing programs relieve individuals of complex decisions that must be made when installing an on-site system. Community solar and wind programs enable participants to invest in a renewable energy project with one phone call or an online sign up. Green power challenges can also make it easier for participants by providing greater exposure to renewable options and, in some cases, providing alternative methods of signing up (e.g., at a farmers' market).

Make a larger impact. For communities looking to transition to renewable energy, many of the mechanisms discussed can provide solutions. Cincinnati recently developed a CCA program; Office of Environmental Quality Director Larry Falkin explained that CCA is one of the biggest opportunities to reduce Cincinnati's carbon footprint (Simes 2012). Green power challenges and reverse auctions also have the opportunity for a large impact.

[1] Opt-out programs are those where customers are automatically enrolled and must take action if they do not want to participate.

Methods for voluntarily procuring renewable energy are detailed in the *Guide to Purchasing Green Power*,[1] which describes how entities can procure renewable energy from utilities or electricity suppliers and in the form of unbundled RECs.

Photo by Dennis Schroeder, NREL/PIX 20373.

A number of new models have emerged through which participants have been able to leverage their purchasing power or work together to lower costs and increase access to renewable energy. This guide focuses on experience with five innovative options currently being used to procure renewable electricity or on-site systems. These next-generation methods do not provide an exhaustive list. Utilities, local governments, businesses, and others are developing new models that respond to market drivers.

2. Community Choice Aggregation

CCA allows communities to determine their electricity generation sources by aggregating the community load and purchasing electricity from an alternate electricity supplier while still receiving transmission and distribution service from their existing provider. CCAs are sometimes described as a hybrid between services offered exclusively by investor-owned utilities (IOUs) and municipal utilities (see Figure 2). CCAs allow for local control of the power source, like in the traditional municipal model, without having to manage transmission lines or provide customer service. CCA programs can choose any type of electricity supply.[2] In the 1990s, communities in Ohio chose alternative suppliers that did not provide renewable energy. However, incorporating renewable energy has been of interest to many communities pursuing CCA in recent years. Other drivers for communities to participate in CCA programs are reduced electricity cost and increasing the percentage of electricity derived from local sources. For the most part, a CCA will purchase electricity from existing renewable energy projects; however, it can be designed to stimulate new development. Cape Light Compact helped establish the Cape & Vineyard Cooperative, which is developing 18.2 MW of new solar at up to 10 local sites. Marin Clean Energy has 31 MW of new, local solar development under contract (LEAN Energy U.S. 2012).

Importantly, implementing a CCA program requires authority from the state. Some states passed CCA-enabling legislation as part of electric restructuring in the late 1990s and early 2000s. Illinois was the state to most recently pass legislation enabling CCAs, in 2009 (see Table 2). We highlight recent experience with this model in this guide due to the resurgence of interest in the mechanism and because of its potential to enable communities to leverage their purchasing power to procure renewable energy.

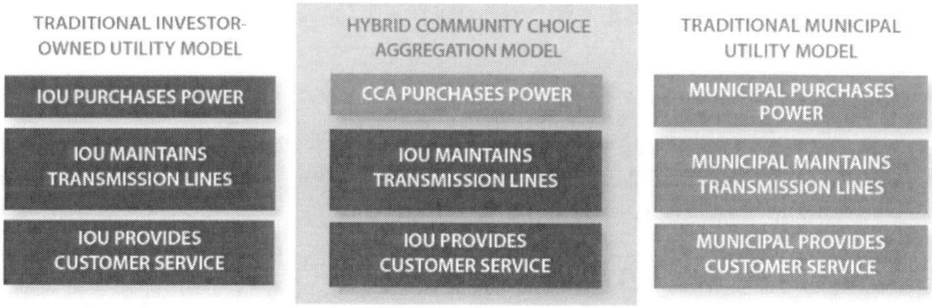

Source: LEAN Energy U.S. (2012).

Figure 2. Investor-owned utility, Community Choice Aggregation, and municipal electric-sector models.

Table 2. States with CCA-Enabling Legislation

State	Year CCA-enabling legislation passed
Massachusetts	1997 (HB 5117)
Ohio	1999 (SB 3)
Rhode Island	2002 (H 7786)
California	2002 (AB 117)
New Jersey	2003 (P.L. 2003, CH 24)
Illinois	2009 (HB 362)

2.1. Experience with CCA

Table 3 provides an overview of CCA programs offering or planning to offer renewable energy options as of March 2012. Illinois is currently a hotbed of CCA development, though additional communities in California and Massachusetts are also pursuing programs.

In "opt-out" programs, all eligible customers are enrolled automatically and it is up to the customer to contact the CCA program if they wish to be removed.

In 2002, Cape Light Compact in Massachusetts was the first CCA in the nation to offer a green option. It is currently offered at a price premium of $0.009/kWh for 50% green power and $0.016/kWh for 100% green power. The program requires customers to opt-in (i.e., customers must take action to sign up), and it had approximately 1,113 green power accounts as of March 2012 for total annual sales of 7,318 MWh of green power. Three additional Massachusetts communities have submitted CCA implementation plans. One of those, the Hampshire Council of Governments, is planning to offer a renewable energy component.

California is soon to have two CCA programs. Marin Clean Energy (MCE) began operating in 2010, offering the Light Green option (50% renewable) and the Deep Green option (100% renewable). MCE had approximately 8,000 accounts as of May 2011 and is expected to reach 100,000 accounts by the time the program is fully rolled out in July 2013. Further details are provided in Text Box 2.

CleanPowerSF is preparing to launch in 2012, providing a 100% green power option. The San Francisco Public Utilities Commission voted in December 2011 to move forward with the program, but the board of supervisors will also need to approve the plan. CleanPowerSF estimated that residential customers would pay an additional $7.00 to $54.50 per month for 100% green power, depending on electricity usage.

Additional communities in California are exploring CCA options. Sonoma County is moving forward with a CCA program, and the East Bay communities of Berkeley and Richmond, in conjunction with the East Bay Municipal Utility District, which is a water agency with 350,000 accounts, are exploring CCA options (Matson 2012).

Table 3. Overview of CCA Programs Offering Renewable Energy

Location	Program Name	Percent of Renewable Energy Content in CCA Product	Type of Renewables	Start Date	Premium	Number of Electricity Customer Accounts	Estimated Annual Sales of Renewable Energy (MWh)
Marin County, CA	Marin Clean Energy	50% or 100% green power	Wind, solar, biomass	2010	100% is $0.01/kWh extra	>8,000	291,000 MWh(2013)
San Francisco, CA	CleanPowerSF	Negotiating for 100% green power	Under negotiation	Expected in 2012	Finalizing terms, conditions, and rates	Estimated at 75,000*	Not available
Cape Cod, MA	Cape Light Compact	50% or 100% green power	Run-of-river hydro, landfill gas, wind, solar	2002	$0.009/kWh–$0.016/kWh premium	1,113**	7,318 MWh**
Cincinnati, OH	Cincinnati aggregation program	100% green power	Partially from local solar projects	2012	Average residential annual savings of $133	52,400 eligible	Not available
Oak Park, IL	Oak Park electricity aggregation program	100% green power	Wind	2012	25% discount to standard supply	~20,000	171,000 MWh

* Sabatini (2011a) ** Soares (2012)

Text Box 2. Marin Clean Energy

Using legislation passed by California in 2002 (AB 117) as a guide, residents of Marin County, California, administered a survey in 2007 in order to gauge local public interest in increasing the community's use of renewable energy. The response to that survey was overwhelmingly positive:

- 90% of residents said that reducing greenhouse gases was important to them, and 74% said they would support the local government becoming a provider of greener energy.
- 69% of residents said they would pay up to 5% more, while 58% said they would pay up to 10% more for an increase in renewable energy (MCE 2012).

The Marin Energy Authority (MEA) was established as a public agency in December 2008 with the target of "significant greenhouse gas emissions reductions" (MEA 2011). In May 2010, the MEA launched MCE.

Based on the market insight provided by the 2007 public survey, MCE developed two levels of green power procurement: Light Green, which provides 50% renewable energy at prices competitive with traditional energy generation, and Deep Green, which provides 100% renewable energy at a price increase of $0.01/kWh. MCE has retained Shell Energy North America as their renewable energy provider, and Pacific Gas & Electric will remain the transmission and distribution provider.

Details within AB 117 provided a key piece to the MCE program success: Customers are automatically enrolled in the CCA unless they explicitly opt out of the program during the initial 60-day period following the commencement of service. MCE achieved roughly an 80% rate of participation in the program's initial phase of implementation. In May 2011, the number of MCE accounts totaled over 8,000. MCE expects similar rates of participation going forward into Phases II and III, due to be complete in August 2012 and July 2013, respectively. MCE expects to reach a total of roughly 100,000 accounts by the end of Phase III. MCE's ultimate goal is to meet the demand of these customer accounts by acquiring 100% of their energy supply from renewable energy sources (MEA 2011).

> Through a combination of demand-side efficiency incentives, a two-tiered subscription system catering to cost-sensitive and environmentally progressive customers alike, a "do nothing" program opt-in strategy, high levels of public support, and favorable net-metering conditions, the MEA expects to achieve a 33% renewable energy supply by 2015. The achievement of that goal would equal California's renewable portfolio standard (RPS) 5 years ahead of schedule.

While all of those communities may not adopt CCAs or select renewable energy, there is large potential in Illinois for CCA-enabled renewable energy purchasing. Potential programs in Illinois have not faced opposition from incumbent suppliers as communities in California have, perhaps because the Illinois electricity market is already deregulated.

Oak Park, Illinois, began its CCA in 2012, providing a 100% wind product sourced from within the state at a price discount of 25% to standard supply. Prices for distribution and transmission remain the same. With approximately 20,000 accounts, the CCA has total annual renewable sales of 171,000 MWh.

Evanston and Peoria, Illinois, have selected renewable energy suppliers, though program designs are being finalized. Evanston has contracted to provide a 100% renewable energy product at a rate about 38% lower than the city's current electric rate.

The future price difference between CCAs and existing supply in Illinois remains to be seen. Existing supply contracts were entered into when power prices were higher, but these are set to expire in June 2013 (Lydersen 2012).

In Ohio, Cincinnati has launched its CCA program for residents and small businesses. The city selected a supplier and is currently negotiating the contract.

2.2. Design and Implementation Considerations

CCA programs are designed by local governments (often with assistance from consultants) within any confines of CCA-enabling legislation.3 Programs differ in terms of the type of renewable offer, price premium, eligible customer classes, opt-out provisions, and the type of renewables selected.

Opt-out provisions. CCAs can be designed either as "opt-in" or "opt-out" programs. With opt-out programs, all eligible customers are enrolled automatically, and it is up to the customer to contact the CCA program if they wish to be removed. State CCA laws may specify conditions for whether programs are opt-in or opt-out. For example, in Illinois, programs can be designed either way, but opt-out programs require approval through a referendum in a general election.

Opt-out rates (the percent of eligible customers that decide to remain with the incumbent supplier) are typically less than 20%. In Marin County, which was heavily targeted with anti-CCA messaging, the opt-out rate was around 20%. In Oak Park, the opt-out rate has been around 3% to 4%; an additional 5% to 6% of residents are not eligible to participate for a variety of reasons. The opt-in rate in Cape Light Compact has been 1,113 customers out of approximately 150,000 customers—or less than 1%.

> A CCA "is probably the biggest opportunity we'll have over the next several years to dramatically reduce Cincinnati's carbon footprint."
> — Larry Falkin, Cincinnati Office of Environment Quality, Director

Type of offer. CCA programs can be designed to offer different percentages of renewable energy. While all existing programs offer a 100% renewable energy product, some programs include other percentage offers. The MEA and Cape Light Compact both offer 50% and 100% renewable energy products. Oak Park only offers the 100% renewable product because the rate for 100% renewables was comparable to the existing utility rate for traditional power.

Price premium. The price premium indicates how much a CCA product costs relative to standard supply. Premiums in current programs vary from zero (Oak Park) to $0.01–$0.02/kWh (Cape Light Compact). As discussed later, some CCA programs are able to offer products at lower cost than standard supply.

Eligible customer classes. Due to differences in state regulation, not all customer classes (residential, commercial, and industrial) are eligible to participate in all CCA programs. In Illinois, industrial and large commercial classes can choose their supplier, without the existence of a CCA. Therefore, Oak Park's CCA is only open to residential and small commercial customers.

Type of renewables. CCA programs can determine the type of energy they wish to supply. Communities may want to support local resources or may favor one technology type (e.g., wind) over another. Oak Park is using 100% wind from Illinois under a 2-year contract with Integrys Energy Services. Marin County uses wind, solar, and biomass, and Cape Light Compact uses run-of-river hydro, landfill gas, wind, and solar. In Oak Park, the city is considering a longer-term (5-year) contract with a wind developer to help support the construction of new wind projects.

2.3. Benefits of CCA

CCA can provide lower-cost renewable energy products, make a large environmental impact, provide an easy way for consumers to support renewable energy, and access lower-cost, tax-exempt debt by using the borrowing capacity of the local government.

Reduced-cost renewable energy. Through aggregation, communities can solicit renewable energy supply at a potentially lower cost than individuals could solicit on their own. A main driver of CCA adoption nationally has been the reduced electricity rates resulting from competition between suppliers.

The City of Oak Park obtained rates for 100% renewable energy that were at a discount to standard electricity rates. In Illinois, communities receive renewable power for an average of 25% less than standard power. In Massachusetts, there is a small premium for 100% renewable energy, with the default supply being 6% cheaper on average. MEA offers competitive rates for partially renewable products, while there is a small premium for 100% products (Marshall 2012).

Large environmental impact at the community level. Many communities are striving to meet already established carbon reduction and/or renewable energy targets. Given that electricity consumption makes up a large part of a city's carbon footprint, developing a renewable offer can produce substantial carbon savings. According to Dawn Weisz of MEA, "When we were reviewing the GHG reductions that would come from various measures we realized that all were very small compared to a CCA, which would take over choosing power sources. With one fell swoop you could eliminate all GHG from power sources" (Braly 2011). In Cincinnati, Office of Environmental Quality Director Larry Falkin explained that approximately 85% of the city's energy currently comes from coal, and that a CCA "is probably the biggest opportunity we'll have over the next several years to dramatically reduce Cincinnati's carbon footprint" (Simes 2012).

Easy for consumers to procure renewable energy. In opt-out CCA programs, no action is required on behalf of the consumer in order to choose renewable energy. This results in larger usage of green power; opt-out CCA programs can have participation rates of more than 80%. Traditional utility green pricing programs (which are opt-in programs) have average participation rates of only around 2%, with the top programs reaching 5.3%–21.5%.

Access to low-cost, tax-exempt debt. CCAs may contract for supply or develop and own supply-side resources. MCE has contracted with Shell Energy for supply but is also developing local renewable energy projects as part of their supply mix. In order to obtain electricity supply, public entities can use low-cost, tax-exempt debt (Speer 2011), while IOUs have a higher cost of capital in order to cover debt, equity, and income taxes. In 2008, MCE estimated its cost of capital conservatively to be in the range of 5.5% to 7.0%, while noting that the incumbent supplier, Pacific Gas & Electric (PG&E), had a cost of capital between 12% and 13% (MCE 2008).

2.4. Challenges of CCA

CCAs require enabling legislation, can potentially have significant start-up costs, may face resistance from the incumbent electricity supplier, and require education and outreach.

Enacting CCA-enabling legislation. In order for a community to establish a CCA, the state must have established a CCA mechanism through legislation. CCAs change the nature of the existing supplier—in regulated states, it challenges the monopoly granted to incumbent utilities, and in deregulated states, CCA changes the role of retail suppliers.

Start-up costs can be significant. Start-up activities include surveying consumers for potential interest, developing an implementation plan, educating consumers about new options, and soliciting offers for supply. Start-up costs for a CCA program can be significant, particularly for communities that are pioneering the effort in their state. In San Francisco, planning efforts for the CleanPowerSF program cost almost $3 million. Start-up costs, including a $15 million escrow account in case the CCA was terminated before contract expiration, were expected to cost $19.5 million (Sabatini 2011b). In Oak Park, the city worked with the Galvin Electricity Institute to develop a program. City staff prepared educational campaigns and relied on environmental groups to campaign for the CCA initiative passage. Costs for future CCA development should be lower than those paid by the pioneering communities.

Possible resistance from incumbent electricity supplier. In California, MCE faced strong opposition from the incumbent electricity supplier, PG&E. In May 2010, the California Public Utilities Commission (CPUC) notified PG&E to cease the use of unapproved opt-out mechanisms that were in violation of tariffs and rules (CPUC 2010a). In a separate docket, the CPUC refined marketing rules for CCA, ordering that utilities would be subject to penalties for marketing or advertising that is untrue or misleading (CPUC 2010b). In Oak Park, however, there was no opposition from the incumbent electricity supplier, likely due to the fact that retail electricity is already deregulated in Illinois.

Need for education and outreach. In Oak Park, the biggest challenge was getting people to think about where their electricity comes from. In the run-up to the city's referendum for approval to pursue a CCA, the city was limited to citizen education—it could not campaign for or against the referendum. Oak Park organized forums, created an energy committee, and partnered with environmental groups to educate local residents.

3. Community Wind and Solar Programs

Community renewables programs provide opportunities for individuals to participate and benefit from larger-scale renewable energy projects located in their vicinity. To date the focus of community renewables has been on wind and solar projects, although the models could be used to develop other types of renewable energy technologies as well. Several ownership models have evolved for community wind projects, some of which are detailed below. While many definitions of community solar programs exist, this guide focuses on community solar programs organized or hosted by a utility company. The emerging crowdfunding model is discussed in Text Box 3.

Text Box 3. Crowdfunding Renewable Energy

Crowdfunding renewable projects differs slightly from community solar or wind projects in that crowdfunding participants provide upfront capital (as a loan) to support the development of the project rather than purchasing shares of the project. Crowdfunded programs allow anyone, regardless of utility territory, to invest in the development of a renewable project. Crowdfunding is used to finance many types of projects, not just renewable energy. Kiva and Kickstarter, for example, are two platforms through which individuals can support a wide variety of projects.

Solar Mosaic (www.solarmosaic.com) is a crowdfunding program based in California, specifically for solar development. It gives anyone (regardless of utility territory) the opportunity to invest in the development of a solar facility, which is typically hosted by a non-profit organization. To date, Solar Mosaic has financed five solar facilities totaling 73 kW by aggregating investments of over 400 people, for a total of more than $350,000. Solar Mosaic offers a full return of each individual's principal over 10 years. In April 2011, Solar Mosaic filed with the Securities and Exchange Commission (SEC) and several states to offer "Solar Power Notes" that would offer a return on investment (Revkin 2012). Details are not yet available on how the return on investment would function. The JOBS Act, signed in April 2012 by President Obama, grants crowdfunded projects raising up to $1 million annually an exemption from SEC securities regulation, allowing them to provide a return on investment, provided that they file initial and periodic disclosures to the SEC.

3.1. Experience with Community Wind and Solar Programs

Cooperative ownership schemes, in which individuals in a community own a portion of the project and receive the associated economic benefits, have been a successful development model for wind energy projects in Europe for decades. Similar models for both wind and solar are now growing in popularity across the United States.

Community wind and solar programs (also known as "solar gardens") allow customers to invest in a wind project or a solar system and receive some of the benefits of the project's production in return, such as tax incentives, utility bill credits, and production incentive payments. A local utility, business, school, non-profit organization, or group of citizens may initiate a community renewables program.

Although community wind and solar programs often offer the project shares to residents and businesses that are close to the project site, the community model can be modified in several ways. The share offer can be initially made to those in the immediate vicinity of the project and then opened to a broader population later if all the shares are not sold. Alternatively, the "community" to which the share offer is made may not be defined by geography. For example, shares may be offered to all members of a specific environmental organization or other interest group, regardless of the members' locations.

Community renewables project ownership can be structured in a variety of ways.

- The 18 investors in the 80-kW Holy Cross Energy photovoltaic project in El Jebel, Colorado, purchased shares at an upfront cost of \$3.15/W. In return, they receive \$0.11/kWh, according to the system's production and the number of shares purchased. The credit appears directly on their monthly electricity bill (Green Power Network 2011a).
- New Centennial Power, LLC, a company formed and jointly owned by community members, is developing the 9-MW Huerfano River Wind Project near Walsenburg, Colorado. Joint owners receive an annual rate of return on their investment, based on the company's profits (New Centennial Power 2012).
- The "Minnesota Flip" structure helps community wind projects take full advantage of the national investment tax credit. A limited liability company (LLC) is formed to own and operate the wind project. The LLC is owned in partnership by local investors and a large equity investor that has greater tax appetite. For the initial years of the project, the equity investor owns the majority of the project, receives investment tax credits, and makes management payments to the local investors. After the project has received all the tax credits it can produce, or the equity investor has received its pre-agreed internal rate of return, the ownership flips such that local investors own the majority of the project and receive the majority of net revenue distributions over the rest of the project's remaining lifetime. The Minnesota Flip structure is a broadly replicable and increasingly

popular ownership model (The Minnesota Project 2009; Windustry 2012).

Community wind was initially spurred by the passage of Minnesota's Community Based Energy Development, or "C-BED", legislation in 2005. Other states, including Illinois, Iowa, Massachusetts, Nebraska, Oregon, Washington, and Wisconsin, have followed Minnesota's lead and adopted their own C-BED legislation.

Community wind and solar programs allow customers to invest in a wind project or a solar system and receive some of the benefits of the project's production in return.

Some key elements of the Minnesota legislation ensure that the benefits of the project are spread broadly through the community to local individuals. Each C-BED project must obtain a local resolution of support. Owners of the projects must be local, and ownership rules limit the percentage that any individual may have in the project. The power purchase agreements of C-BED projects must provide levelized cash flow to the project owners. In addition, public utilities are required to consider C-BED projects when they make new energy acquisitions and to set a special tariff for C-BED projects, based on net present value. Utilities are also encouraged to assess interconnection issues for C-BED projects.

Community solar programs appeared in the United States around the same time as community wind but were somewhat slower to take off, perhaps due to higher technology costs. In 2011 there was substantial growth in community solar projects, however, with 10 new programs developed that year alone (see Figure 3). Overall, the number of programs has increased from 3 in 2008 to nearly 30 projected by the end of 2012. The appendix provides a list of community solar programs and relevant program details.

In 2011, Salt River Project, a utility in Arizona, established the largest community solar program on a capacity basis with the launch of its community solar program. Salt River Project offers blocks in a 20-MW solar facility, of which 7.8 MW are designated for purchase by schools and 2.0 MW are designated for residential purchase (Green Power Network 2011b).

Community members can be included in relevant siting decisions; local businesses can be used in construction and maintenance activities;

and local schools may benefit from educational opportunities offered community renewables projects.

Some community solar programs have expanded over time, indicating the popularity of the model. For example, Ellensburg, Washington, began with a 27-kW system in 2006 and expanded the program to 111 kW. In Colorado, the Clean Energy Collective, working with Holy Cross Energy, expanded its program from 80 kW in 2010 with a new 858-kW system installed in 2011.

3.2. Design and Implementation Considerations

In developing community-funded programs, the following are important design and implementation issues:

Program size. Community solar and wind programs can be developed regardless of the size of the system. Some programs are just a few kilowatts (Bainbridge Island's 5-kW program), while others are on the megawatt scale (Salt River Project's 9,840-kW program).

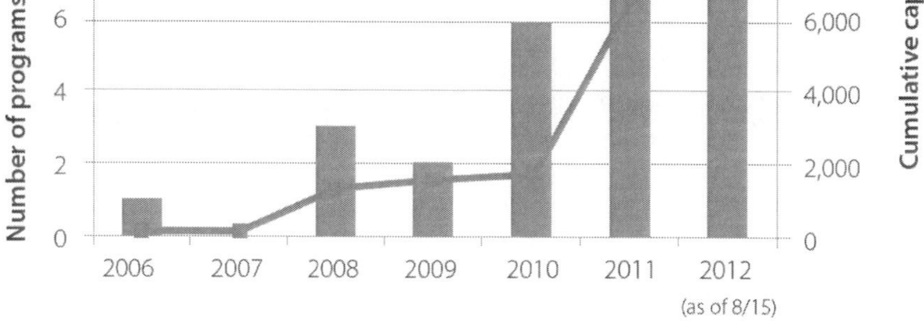

Source: NREL (2012).

Note: Only the residential portion of Salt River Project's community solar facility is included above. An additional 7.8 MW are designated for purchase by schools.

Figure 3. Capacity and number of community solar programs.

Upfront cost. Community solar programs typically allow customers to purchase a panel or unit upfront (sometimes referred to as a "share" or "slice"), though some programs allow customers to purchase smaller increments. Upfront costs for panels typically range from $3–$6/W for a minimum purchase of one panel (around 200 W). A few programs allow customers to purchase smaller increments. Trico Electric, a utility in Arizona, allows customers to purchase quarter panels at $230. Delta-Montrose Electric Association (Colorado) allows customers to purchase blocks of 2.67 W for $10 each. Tucson Electric Power (Arizona) bases the premium on 150-kWh blocks at $3 per block.

Bill credits. Community solar participants generally receive a bill credit per kilowatt-hour based on the electricity production of their share. Rebates per kilowatt-hour typically range from $0.07/kWh to $0.11/kWh. For example, Seattle City Light offers $0.07/kWh generated. In addition, some customers may receive upfront rebates (payments based on the kilowatt purchased) or incentive payments from the state (in Washington, up to $1.08/kWh can be obtained). Colorado Springs Utilities offers a one-time upfront rebate of $1.08/W, up to 30% of the participant's investment.

Community inclusion. For community wind projects in particular, broad community participation in the planning, implementation, and economic benefits of the project can play a role in public acceptance and the ultimate success of the project. For instance, community members can be included in relevant siting decisions; local businesses can be used in construction and maintenance activities; and local schools may benefit from educational opportunities offered by the project. In the United Kingdom, a wind turbine in the town of Swaffham was constructed with a viewing platform under the nacelle, and for a small fee tourists can climb a stairwell up the turbine's tower to get a bird's eye view of the region (EcoTech Centre 2012).

Renewable energy certificate treatment. RECs represent the environmental attributes of 1 MWh of electricity generation. RECs can be sold separately from electricity generation. Without ownership of RECs, environmental claims cannot be made. For example, if a commercial customer were to participate in a community solar program where the RECs stayed with the local utility, the commercial customer could not advertise that it was solar powered. Community solar programs generally, but not always, allocate the RECs to the utility. Similar to many utility solar incentive programs, which provide residential and commercial customers with an upfront rebate to install solar, community renewables projects often ensure that RECs stay with the utility, allowing it to meet any RPS or other environmental target. In theory, the allocation of the REC to the utility should result in a lower purchase cost to the participant, although each case may vary. Participants not receiving RECs may have other reasons for joining community programs, such as supporting local development.

3.3. Benefits of Community Wind and Solar Programs

Community wind and solar programs offer many benefits, including: expanded access to renewables, ease of participation, cost efficiencies, local connections, and local economic benefits. The program design is highly flexible and encourages innovative solutions to local challenges.

Expands access. Community programs allow renters and homeowners with poor solar resource or shading to receive the financial benefits of renewable energy.

Easy to sign up. Residents can be overwhelmed by the process of installing solar on their home. Community programs allow customers to "go solar," or participate in a wind energy project, with one phone call, although project organizers may spend years developing the project.

Cost efficiencies of a larger project or procurement. Depending on the size of the community solar project, there could be cost savings from building one large project rather than many small distributed systems, which should be reflected in the cost to purchase from a community program.

Increase connections and reduce tensions. Wind (and solar) projects that include local ownership and involvement are more likely to be publicly accepted. Community renewables programs help make local connections between the variety of stakeholders involved. Positive models encourage community discussion and involvement in the earliest stages of the project. These projects can be a source of community-building and pride within a locality, in addition to the economic and environmental benefits they provide.

Increased economic development impacts. Research has shown that community projects provide more jobs and increased economic impact over traditional energy projects. A U.S. Government Accountability Office study found that local ownership of wind systems generates an average of 2.3 times more jobs and 3.1 times more local dollar impact compared to other projects. Overall, such studies indicate that community-based wind projects provide around 1.5 to 3.4 times the economic benefit to communities during the operation of the project than projects owned by absentee companies (Lantz and Tegen 2009).

Flexibility. Community renewables programs may be initiated by utilities, existing groups, and/or a few local citizens. The programs may define "community" in a variety of ways and do not have to be dependent on a participant's physical location. Community solar membership is often tied to the sponsoring utility's service territory, but if participants move they typically can keep their membership (if they move within the utility service territory) or sell the membership (if they move outside of the utility service territory). A variety of mechanisms for return on investments can be used.

Encourage innovative ideas. Community projects often spawn creative solutions to challenges. Community renewables projects have been used to test innovative financing models and as a way for equipment manufacturers to enter new markets with which they have limited experience (Bolinger 2011).

3.4. Challenges of Community Wind and Solar Programs

Community programs certainly face challenges. Policy favors large-scale investors/developers, programs have long development times, there is a need to fund initial project development costs, and there are restrictions on offering a return on investment.

Policy favors large-scale investors/developers. Existing policy structures often create barriers to the development of community-owned projects. The federal production tax credit, for instance, is designed to make use of the tax appetite of large-scale investors, thus government entities and small investors often cannot take advantage of tax credits. It can also be difficult for community-owned projects to obtain loans through traditional avenues. Public entities may also not be able to make use of options like Clean Renewable Energy Bonds.

Long development times. Community renewables project leaders need to be persistent. They should be prepared for many hurdles and for long development times. David Brosch of University Park Community Solar noted, "It took us over two years to develop our project structure and only two months to find our members" (DOE 2011, p. 34). Some community wind projects have experienced development times of 5 years or more (Bolinger 2011; The Minnesota Project 2009).

Need to fund initial project development costs. Those planning community renewables projects often focus their efforts on raising funds for construction and permanent financing costs; however, many such projects will incur significant costs before even getting to the construction financing stage. Raising the seed capital needed to cover the initial costs of developing community renewables projects can be a significant hurdle for some community projects (Bolinger 2011).

Restrictions on offering a return on investment. Community renewables programs should be aware of potential securities regulation. In order to avoid any appearance of selling securities, the DOE's *A Guide to Community Solar* recommends avoiding "references to 'shares' or 'stock,' since those terms are the classic ones used to describe securities issued by a corporation and might create an expectation of profits and other rights customarily associated with stock or shares" (DOE 2011, p. 33).[4]

4. Green Power Challenges and Local Collaborative Electricity Procurement

Communities can encourage members to purchase voluntary green power by developing a green power challenge. In many cases, communities partner with local utilities, third-party marketers, and/or environmental organizations to raise awareness of green power purchasing options. With buy-in from a large stakeholder group, community challenges can increase participation in utility green pricing programs by providing multiple communication avenues. Short-term green power challenges set goals for having a certain fraction of their residents and businesses purchase green power within a specified time, usually around 6 months.

Local collaborative electricity procurement aggregates the demand of a few organizations in deregulated electricity markets to procure renewable electricity on a voluntary basis. This mechanism is similar to CCA but only involves a limited number of organizations, which must be located in states with retail choice. Because retail choice states allow customers to switch electricity suppliers, customers can band together to negotiate a better price or product. This type of procurement has happened in the Washington, D.C., area through the assistance of a non-profit organization, Groundswell. Groundswell does not supply electricity but rather coordinates large customer procurement in retail choice states.

Though not developed as "challenges" or campaigns, other initiatives have sought to highlight the ability to purchase renewable energy in a given state. The ChoosePAWind initiative provides links to retail electricity suppliers and REC marketers sourcing supply from Pennsylvania wind facilities. The initiative also highlights the economic benefit of purchasing from local renewable facilities. A similar service focused on New York, Green Power NYC, provides links to competitive suppliers sourcing renewable energy from New York State facilities. The effort is sponsored by the Natural Resources Defense Council and the Alliance for Clean Energy New York.

4.1. Experience with Green Power Community Challenges and Local Collaborative Electricity Procurement

Green power challenges proliferated in the late 2000s as a way to boost involvement in utility green pricing programs. More recently, local collaborative electricity procurement has emerged as a way for organizations to buy renewable electricity at lower cost.

Green Power Challenges

Challenges were particularly active in Oregon, where Portland General Electric (PGE) worked with numerous communities to develop and implement challenges. In 2006, PGE worked with the City of Salem, Oregon, which ran a challenge for 6 months with the goal of attracting upwards of 500 customers. Since then, PGE has also worked with the City of Beaverton, the City of Gresham, and the City of Lake Oswego (see Text Box 4).

Text Box 4. Green Power Challenge in Lake Oswego, Oregon

In 2009, Lake Oswego, Oregon, the sixth-largest city in the Portland area with a population of 36,000, launched a 2-month Green Power Challenge. Lake Oswego had been looking at sustainability issues for the past 10 years, and in 2007 adopted a sustainability plan that, among other things, examined energy use. PGE approached the city's newly formed Sustainability Advisory Board with a Green Power Challenge proposal, and the board recommended that the city council support the proposal.

Working with PGE, Lake Oswego set a goal of 300 new participants. The mayor of Lake Oswego began the Green Power Challenge by issuing a proclamation. The city promoted the challenge through its monthly newsletter, press releases, and website. Working with PGE, they conducted door-to-door outreach and tabled at the local farmers' market.

Ultimately, Lake Oswego exceeded its goal, enrolling 336 new residential and 20 new business customers in PGE's renewable programs. The community benefited by receiving national recognition. EPA staff made a presentation to the city council, street signs promoting Lake Oswego as a Green Power Community were installed, and the city received favorable local press coverage. As of February 2010, more than 1,600 residences and businesses enrolled in one of PGE's programs and the community was collectively purchasing 9% of its power as green power, or almost 35 million kWh of renewable power.

Each of these cities, along with 28 others, are also part of U.S. EPA's Green Power Communities program, which provides public recognition to communities for meeting or exceeding minimum renewable energy purchase requirements. Purchase requirements vary depending on the community's annual electricity usage (Table 4). Since the program launched, the purchasing by the top community has increased from 163 million kWh in September 2010 (Santa Clara, California) to 752 million kWh in June 2012 (Washington, D.C.). Table 5 provides a list of the top five EPA Green Power Communities in terms of annual green power usage and the green power percent of total electricity use. Green power sales in these communities are heavily dominated by large purchases by non-residential participants. Communities can play a large role in the voluntary green power market; EPA's Green Power Communities are currently collectively buying more than 4.2 million MWh of green power annually, which represents 12% of all voluntary sales in 2010.

Local Collaborative Electricity Procurement Model

The local collaborative purchasing model aggregates purchasing power from existing community organizations in retail choice states to solicit competitive bids for renewable electricity generation. Instead of purchasing shares of a renewable facility or providing upfront capital, this model seeks to procure renewable electricity from one alternative supplier. By aggregating the electricity demand of multiple participants, cost savings can be achieved. Groundswell, a Washington, D.C.-based non-profit organization, operates the Community Power Project, which works with neighborhood organizations and faith institutions. For example, First Trinity Lutheran Church partnered with other Washington, D.C., churches to buy wind power and saved $6,000 annually. Groundswell estimates that since 2011 the Community Power Project has procured $5 million in renewable electricity, with participants saving up to 20% annually on energy bills (Groundswell 2012). Groundswell has organized 109 organizations to procure 56,000 MWh of renewable energy, the majority of which (98%) was Green-e Energy Certified wind (Witherbee 2012).

Table 4. Green Power Community Purchase Requirements

Community Annual Electricity Usage	Minimum Green Power Community Purchase Requirements
>100,000 MWh	3%
10,001–100,000 MWh	5%
1,001–10,000 MWh	10%
≤1,000 MWh	20%

Source: U.S. EPA (2012a).

Table 5. Top EPA Green Power Communities (as of June 19, 2012)

Rank	Community	Annual Green Power Usage (MWh)
1	Washington, D.C.	752,505
2	Portland, OR	708,667
3	Hillsboro, OR	678,600
4	Philadelphia, PA	593,309
5	Aurora, IL	250,975
1	Oak Park, IL	91.9%
2	Brookeville, MD	45.7%
3	Hillsboro, OR	33.7%
4	Swarthmore, PA	27.9%
5	Corvallis, OR	21.2%

Source: U.S. EPA (2012b).

Although the Community Power Project operates in retail choice market and procures bundled renewable electricity (energy and RECs), this model could be used for unbundled REC purchases as well.

4.2. Design and Implementation Considerations

Green power challenges and local collaborative electricity procurement programs need to set appropriate, realistic targets, decide how to distribute costs among partner organizations, and consider administrative issues.

Setting appropriate, realistic targets. With green power challenges, communities can work with their local utility to establish a target of either number of participants or percentage renewable energy purchased. Utilities can help partners understand existing participation and sales rates. Final targets for a percentage of sales from renewable energy are already established for communities seeking to join the EPA Green Power Communities program; working with the local utility company can help partners develop a plan to reach these targets.

Distributing costs among partner organizations. Organizations should determine in advance who will pay for the various components of the program.

Considering administrative issues. Consider what information is needed in order for a consumer to sign up for a green product and work with the utility to streamline the enrollment process. For example, if consumers need to provide their electric account number, this may prohibit them from signing up at community events like farmers' markets and concerts.

4.3. Benefits and Challenges of Green Power Challenges and Local Collaborative Electricity Procurement

Community green power challenges can contribute to the branding of a community, provide education and outreach to community members, and provide an easy way for community members to support green power.

Raise awareness and expand access to renewable options. A large barrier to participation in utility green power programs is lack of awareness.

According to the Natural Marketing Institute, approximately one in six adults are aware of renewable power options (NMI 2011), yet more than half of consumers have the option to purchase renewable energy through their utility (Bird and Sumner 2010). Of course, all consumers have the ability to buy RECs separately from their electricity. Community challenges can help overcome the awareness barrier by targeting community residents during short, intense campaigns using multiple methods of communication. Education efforts by non-profits seeking to aggregate demand can also help organizations better understand renewable energy options.

Easy to sign up. By providing different ways of enrolling (e.g., at a farmers' market), green power challenges can make it easier for consumers to sign up.

May encourage friendly competition among suppliers. In regulated electricity markets, utilities offer green pricing programs, while in deregulated or restructured states, competitive marketers offer renewable options. In all states, renewable energy can be supplied by RECs independently from electricity. By running a campaign that includes a utility or competitive marketer in addition to a REC supplier, communities may encourage friendly competition. Boulder's Wind Power Challenge campaign worked with the local utility, Xcel Energy, and REC suppliers Clean and Green, Community Energy, and Renewable Choice Energy. Xcel Energy's green pricing program is in competition with REC suppliers for customers. Local collaborative electricity procurement also focuses on getting the lowest-cost product from suppliers.

Contribute to branding of a community. Green power challenges generate publicity for the local government and any other co-sponsors, such as large institutional purchasers and the local electric utility. In Oregon, communities looking to attract high-tech businesses were interested in promoting their community's green image (Hinckley 2010).

Green power challenges and local collaborative electricity procurement programs need to set appropriate, realistic targets, decide how to distribute costs among partner organizations, and consider administrative issues.

Lower cost. By aggregating demand and soliciting renewable electricity bids for multiple organizations, local collaborative electricity purchasing through Groundswell achieved up to 20% annual savings on energy bills for participants (Groundswell 2012).

Green power challenges are fairly straightforward to implement. The primary implementation activities may include scheduling the challenge with the city council, determining the timing, and getting space at the local farmers' market. Because many organizations might be involved with a green power challenge, it is important to clarify funding needs upfront and determine which organizations will cover which costs. Local collaborative electricity procurement requires a lead organization to educate potential partners and develop the actual procurement.

5. Bulk Purchasing of Distributed Energy Systems

Bulk purchasing programs are an increasingly common mechanism to encourage solar installations, in particular, but could also be used for other distributed energy technologies (U.S. Department of Energy 2011a). These programs identify a group of individuals or companies interested in installing solar systems and aggregate their demand. By buying the systems in bulk, group purchasing attracts reduced prices from vendors. The programs also save buyers the time and effort that would be required to navigate the purchasing process individually.

A bulk purchasing program requires a clear leader to organize the process. The leader may be a local government agency, interest group, cooperative, private enterprise, or a partnership between these. The organizer plays a variety of important roles throughout the process, including:

- Defining the potential participant base
- Seeking out participants
- Providing information and education on renewable energy options
- Assessing the possible sites and types of demand and conducting resource assessments
- Issuing a request for proposals (RFP) to potential vendors
- Selecting the winning vendor(s)
- Overseeing the signing of standard contracts between participants and vendors
- Assisting with coordination during the installation phase.

5.1. Experience with Bulk Purchasing of Distributed Energy Systems

The group purchasing model began in the United States with the Solarize Portland initiative, which functions as partnerships between the city and neighborhood associations. The model has since grown to include a wide variety of programs that are organized by not-for-profits, private enterprises, public utility commissions, and governmental agencies at various levels. Examples of several of these models are provided below.

City Government Model

Portland, Oregon

Neighborhood associations in the City of Portland can obtain technical and programmatic assistance from the city to offer group solar purchasing programs to their residents. The city assists the associations in program design, provides outreach and educational materials, and aids in the selection of solar contractors. It also helps coordinate and deliver educational workshops about the volume purchasing concepts and the benefits of solar energy. The Energy Trust of Oregon, Solar Oregon, and local solar contractors also played key roles in the program's success. Since its beginning in 2009, the program has resulted in over 600 solar installations and cost savings of up to 35% (U.S. Department of Energy 2011a). Additional programs have been developed in Pendleton, Beaverton, and Salem, Oregon.

Los Angeles, California

Other cities across the nation have followed Portland's lead. In 2009, a group of citizens in Los Angeles County, California, came together to form the Open Neighborhoods community solar program. Solar panels were installed on 32 homes in the first round. Another round of installations was organized in 2011, which brought the cost of solar close to that of electricity from the grid. The price of around $4.40/W was reported to be the lowest cost for residential solar systems in California. The Open Neighborhoods program encourages broader participation by making donations to schools and non-profits when community members participate in the group-buy program (Farrell 2011).

Santa Barbara, California

The Community Environmental Council of Santa Barbara, California, initiated a pilot group purchasing program in 2011, installing 49 systems in 3 months. The program began its second phase in August 2012, offering both solar system purchasing and lease options to satisfy a variety of financing needs. The council selected two solar contractors to provide the systems, obtaining a bulk price for program participants. A one-time fee of $0.15 per installed watt is paid by the contractors to fund the council's continued efforts.

Madison, Wisconsin

The City of Madison, Wisconsin, helped facilitate the first residential group solar purchasing effort in Wisconsin. The 22 participants in the first round of the MadiSUN program obtained solar systems for an average of $5,320/kW, a discount of 20% (MadiSUN 2012). Marshfield, Wisconsin, has also started a group-buy program, Solarize Marshfield, funded by a DOE SunShot grant (Midwest Renewable Energy Association 2011).

Photo by Dennis Schroeder, NREL/PIX 21499.

Private Enterprise Model

One Block Off the Grid

One Block Off the Grid (1BOG) is a for-profit enterprise that acts as an intermediary between potential customers and solar installers. The company identifies potential solar system customers, provides information on solar options, and finds and vets installers on behalf of the customers, obtaining bulk pricing for the systems. The service of amassing the demand within a region is valuable to system providers as well, who pay a fee to 1BOG for each watt they install for customers through 1BOG. The fee is built into the price of the installed solar systems but is more than offset by the price savings obtained from buying in volume. 1BOG's success began when they assisted the San Francisco Department of the Environment in conducting aggregated solar purchases at the request of several neighborhoods in the city (U.S. Department of Energy 2011b).

Utility Model

Orlando Utilities Commission

In 2011, the Orlando Utilities Commission (OUC), a municipal utility company, carried out the pilot Commercial Customer Solar Aggregation program to assist its large commercial customers with installation of solar technologies. OUC identified nine large commercial customers interested in installing solar systems. OUC issued an RFP for this aggregate demand, which totaled an estimated 1,237 kW of solar electric and 77 kW of solar thermal capacity. Developers were allowed to bid on solar thermal, solar electric, or both. The chosen developer will construct, own, and finance the solar projects through a power purchase agreement (PPA), with OUC acting as the billing agent. OUC purchases the RECs and excess electricity generated at the facilities under a long-term fixed-price PPA of at least 25 years (OUC 2011).

Collaborative Partnership

City of San Jose, Bay Area Climate Collaborative, and San Jose Credit Union

The City of San Jose, the Bay Area Climate Collaborative,[5] and the San Jose Credit Union partnered to form a solar group-buy program called SunShares, supported by a federal Solar America Cities grant. The program offers employees and retirees of the over 360 Silicon Valley Leadership Group member companies the opportunity to purchase solar thermal and photovolatic systems at discounted rates and take advantage of financing options as low as 3.99%. The program is administered with assistance from GroupEnergy, a company specializing in collaborative purchasing programs (BACC 2012).

Successful bulk purchasing programs explore creative financing options to enhance customer participation and encourage competition between many vendors through effective outreach.

Silicon Valley Collaborative Renewable Energy Procurement Project

Another collaborative, the Silicon Valley Collaborative Renewable Energy Procurement Project, was formed through a partnership between the City of Santa Clara, Joint Venture: Silicon Valley Network, and the Public

Sector Climate Task Force.[6] The program facilitates public agencies in the installation of renewable energy systems. Nine public agencies are participating in the first phase to develop 70 sites on 43 locations, for a total capacity of 14.4 MW. As such, the effort represents the largest multi-agency procurement of renewable energy in the United States at this time. The agencies sign a Memorandum of Understanding in the initial stages to express their intent to carry through the entire development process.

Types of sites identified for development include office rooftops, carports, water storage tanks, ground-mounted systems, bus depots, senior centers, parking garages, and health centers. Site assessments were contracted prior to the release of the RFP, both to ensure site feasibility and to provide bidders with accurate site information on which to base their bids. These technical and economic assessments were undertaken by Optony, Inc. for all of the sites, on behalf of the participating agencies. This relieved the agencies of the need to conduct individual analyses, for which they may not have resources or expertise. While this process reduced the number of sites considerably, it ensured that only sites that met technical and economic feasibility were included in the bidding process.

Because the systems were of a broad range of sizes, the project team grouped the installations into strategic bundles of 6–15 locations each, based on system size and across different owners. The team allowed developers to bid on one or more bundles. This encouraged participation by bidders who specialize in a subset of system sizes or who would be deterred if the entire contract were offered to a single bidder. A point system in the selection process gave preference to local vendors.

Participating agencies buy the power generated by the systems from the developer but have the option to purchase the system at several stages in the contract period. This arrangement means that agencies that cannot afford the high upfront cost of going solar can still participate.

According to the Solar Labor Force Impact Model, approximately 200 jobs will be created by the program, with one-third of those being permanent positions in fields such as contract management and system maintenance (U.S. EPA 2010; Joint Venture: Silicon Valley Network 2011).

As a result of the success of the program, the EPA is implementing a similar effort called the Clean Energy Collaborative Procurement Initiative, which targets federal, military, and higher education facilities in the Washington, D.C., metro area (U.S. EPA 2011).

5.2. Design and Implementation Considerations

General

General design and implementation considerations include:

- Consider local resource availability and likely participant base
- Leverage an existing group or organization to serve as the program leader; look for strategic partnerships and make use of available expertise in outreach, financing, site assessment, and contracting
- Identify the state and federal incentives; keep an eye on them and meet the deadlines
- Use independent experts to assess and refine your program plan and to carry out site assessments
- Explore creative financing options to enhance participation
- Identify program goals and metrics of success early in the process.

Communication and Participant Outreach

Communication and participant outreach considerations include:

- Use a variety of communication methods with program participants; avoid a one-size-fits-all method
- Provide contact options that suit a variety of communication styles
- Maintain participant satisfaction with as much personal communication as feasible
- Hold informational meetings to encourage broad participation and to answer participant questions throughout the process
- Create an online forum to inform interested communities, connect participants, and solicit program feedback.

Bidding Process

Program success depends on both buyers and sellers benefiting from participation. In order to accomplish this, programs could:

- Provide information to vendors on the benefits of aggregated demand
- Encourage competition between many vendors through effective outreach
- Bundle sites by installation type, host facility, size, and other attributes

- Allow vendors to bid on one or more of the bundles.

Vendor Selection

The strength of a vendor can impact the success of a program. In selecting a vendor, programs could:

- Consider a variety of factors when selecting winning bidders, such as cost, system design, quality assurance, and ability to provide long-term maintenance and support
- Select one or more reputable vendors with a track record of high-quality installations and be very clear about the terms of the contract.

5.3. Benefits of Bulk Purchasing Programs

Bulk purchasing programs provide a variety of benefits, ranging from lower costs to broader adoption of solar energy.

Lower costs. Aggregating demand unlocks volume discounts from local solar installers.

Reduced transaction costs. Administrative costs are spread amongst a large group of customers.

Reduced risk perception/ease of decision making. As a member of a group, participants have reduced risk perception. The leadership of the organizing group relieves individuals of complex decision-making processes.

Company reputation/employee satisfaction. Companies and agencies can offer group-buy programs as part of their employee or membership benefit packages. The unique offer can increase employee satisfaction, attract new members, and enhance the company's overall reputation for sustainability.

Encourage broader use of solar energy. Because group-buy campaigns can be made available to employees, neighborhoods, and other broad bases of individuals, they encourage people who would not have otherwise considered or investigated the option to install solar energy. Peer pressure to participate may work to increase solar deployment.

Public education. Through workshops and participant outreach, group-purchasing campaigns enhance overall public education about solar energy.

Solar providers. Vendors of solar systems receive substantial new business with reduced sales expenditures.

Competitive contract terms. Developers are more likely to agree to competitive contract terms (e.g., buyout options, performance guarantees, and termination options) when making bulk deals.

Flexibility of design. Group-buy programs can be designed to suit a wide variety of goals and needs. They can incorporate direct purchases, leases, or unique financing and ownership models. They can target a variety of participant types, system sizes, and technologies.

Quality assurance. Thorough vendor-selection processes ensure system quality, which improves buyer satisfaction as well as the overall reputation of solar energy.

5.4. Challenges of Bulk Purchasing Programs

Bulk purchasing programs require a champion and provide less flexibility than an individually designed procurement process.

Requires a champion. One or more organizations must be willing to take the lead and have the resources to carry out the program.

Less flexibility. Individual buyers may have less flexibility to select vendors, customize systems, or specify contract terms.

Scheduling challenges. Participating installers may be pressed to complete many installations in a short time frame. The individual participant is subject to program schedules and deadlines. If individuals are required to wait too long for the installation of their system, they may be tempted to withdraw from the program and accept bids from vendors that are not a part of the program.

Photo by Dennis Schroeder, NREL/PIX 20689.

6. Reverse Auctions for Voluntary Purchases

A reverse auction is an auction in which the roles of the buyer and seller are switched. In a traditional auction, a seller puts a product up for sale, which is bid on by many potential buyers. The buyer willing to pay the highest price wins the auction. In a reverse auction, a buyer requests a product, and many sellers bid to provide that product. The bidder that can provide the product at the lowest price wins the auction.

A few elements are essential to the functioning of a reverse auction:

- The product to be supplied must be a standardized product of known quality
- There must be multiple sellers of the same product
- There must be a range of prices at which the product can be produced and sold.

Energy production is a natural fit for a reverse auction because it meets all three of the above criteria. The standardized product to be supplied is electricity. There are many energy developers who can compete to provide this product, and the price at which power can be provided is variable.

Reverse auctions provide a flexible method to purchase renewable energy generation in deregulated electricity markets. In addition, reverse auctions can be used to purchase RECs or to contract for the construction of a renewable energy system in all electricity markets.

Private companies, government agencies, and non-profit groups can all use a reverse auction.[7] These buyers can either organize and hold their own reverse auctions or use an independent auction manager to coordinate auctions on their behalf. Using an auction manager avoids the need for buyers to go through the learning process needed to organize a successful auction. World Energy Solutions is one company that offers reverse auction services for clients in deregulated electricity markets.

Well-designed auctions create an open, market-driven process that draws in a variety of potential product suppliers. Pre-auction planning and analysis helps buyers and sellers make informed decisions on auction day. Thorough pre-auction documents ensure that all bidders are clear about the product and terms of provision on which they will bid.

Once the auction opens, all bidders see the prices their competitors are willing to contract for. In the moments prior to auction closure, the opportunity to make a final blind bid encourages bidders to submit their lowest possible final offer. The transparency of the process ensures equality and enhances participant trust while creating a highly competitive environment that drives down prices.

Reverse auctions can be used for entities that aggregate their demand to increase their purchasing power. Reverse auctions in which buyers are requesting bids to provide large quantities of electricity draw more bidders and typically result in lower prices per kilowatt-hour than smaller auctions. World Energy Solutions runs auctions for clients with demand levels of 4+ million kWh/year (roughly the demand of 350 American households) (Joyce 2010). Aggregating demand can allow buyers to take advantage of this leveraging effect. Aggregating the demand of several buyers within the same power service area allows those with smaller demands to participate in the reverse auction process. Even aggregating accounts that are already large may provide further price benefits.

Combining the purchase of traditional electricity with that of green energy, such that total demand is supplied through a single auction, is another form of aggregation.

6.1. Experience with Reverse Auctions

The Healthcare Clean Energy Exchange (HCEE) provides an example of how buyers with limited demand can be aggregated to allow for broader participation in reverse auctions. The HCEE is a reverse auction service offered to its members by Practice Greenhealth, a non-profit organization for healthcare facilities. Through HCEE, member facilities can participate in custom-tailored, online reverse auctions to procure renewable energy or the development of hospital-sited renewable energy systems. The exchange makes use of the World Energy platform described above. In 2009, Mercy Hospital and Medical Center participated in their first reverse auction with the goal of reducing their annual electricity expenditures, which totaled about $2.5 million. The facility procured 10% green electricity and saved more than $190,000/year compared to their previous utility contracts (Pizzi 2009). Multiple hospitals in the same distribution area may also choose to aggregate their renewable energy demand in order to reach the capacity minimums required for auction participation and further drive down prices (Practice Greenhealth 2012).

Well-designed auctions create an open, market-driven process that draws in a variety of potential product suppliers.

The General Services Administration (GSA) provides another example of how a reverse auction can be used to procure power for multiple loads with notable price benefits. The GSA is an independent agency of the U.S. government that supports federal agencies by managing assets, such as office buildings and transportation fleets, and leveraging the government's buying power through bulk acquisition of products and services. Since 2001, GSA has used the World Energy Solutions services to procure electricity for government facilities across the country. More recently, GSA has used in-house expertise to hold reverse auctions themselves (Shah 2011).

Through a reverse auction, GSA has obtained electricity for federal agencies at significantly lower prices than existing contracts. Through a round of auctions organized by World Energy in 2008, GSA purchased 100% wind power for the Statue of Liberty and Ellis Island at prices under the current standard offer contract rate. Through a round of auctions in January 2012, new contracts were signed to provide electricity to New York federal facilities for 35% lower cost than previous utility contracts (World Energy 2012).

6.2. Design and Implementation Considerations

A reverse auction can be used to procure renewable energy in deregulated (competitive) electricity markets and can be used to procure RECs in all electricity markets.

Clearly written and thorough pre-auction announcements are an important feature of a successful reverse auction process. These documents specify the amount of electricity to be purchased, any technology or location restrictions, and all contract terms that would be offered to a winning bidder. These documents ensure that bidders understand exactly what they will be bidding on.

A reverse auction can be used to influence the market in a number of ways, including:

- Encouraging the development of new renewable energy facilities by specifying that the generation to be purchased must come from a new facility
- Encouraging local development by specifying that the generation to be purchased must come from within a particular region
- Encouraging a particular technology (e.g., solar) by specifying that the generation to be purchased (or a portion thereof) must come from a certain technology.

Ongoing verification and auditing requirements ensure that the renewable power being procured through the reverse auction is not also being used to satisfy a state RPS or other renewable energy targets.

In a reverse auction, each bidder aims to bid lower than all other bidders. Bidders will only bid as low as necessary to beat the most recent bid. If the auction time runs out before a bidder reaches their lowest possible offer, the buyer could pay more than necessary for the product requested. Thus, a pure reverse auction may not result in the lowest possible price for the buyer. This drawback can at least partially be addressed by switching the reverse auction to a blind auction for the final seconds of bidding. All bidders place their best and final offer at the end of the auction, without seeing their competitors' final bids. This increased uncertainty during the final seconds of the process encourages bidders to place their best offer on the table.

Today's reverse auction platforms are high-speed, time sensitive, and dependent on the Internet. If a bidder has technological difficulties at the time of the auction, their bids will not be available for consideration. The technology requirements may also inhibit some buyers from holding their own reverse auctions and necessitate going through an auction coordinator at additional cost. Some auction platforms require bidders to register for a fee, which could discourage potential providers.

6.3. Benefits of Reverse Auctions

Reverse auctions are not entirely a new concept in the electricity supply realm. Many of the same elements are present in the traditional methods for acquiring electricity supply contracts, such as RFPs or requests for bids. Several benefits, however, set the reverse auction process apart from traditional methods of acquiring electricity service contracts. The benefits of using a reverse auction include:

Increased transparency. Favoritism between bidders is impossible during the auction process, and all bidders have an equal opportunity to win. There is also price transparency with all parties able to see the lowest bid. The impartiality and transparency of the auction process increases confidence amongst bidders and encourages broad participation.

Increased competition. Today's online reverse auctions are fast-paced. The details of the product being requested and the contract terms that will be offered are announced well in advance so that bidders can prepare. The actual bidding process typically lasts for only a few minutes. While the auction is open, bidders receive instant feedback on the prices their competitors are offering. They know that if they do not respond with a lower bid quickly, they will lose the contract. This environment taps the competitive nature of bidders and drives down prices.

Using a reverse auction process, Mercy Hospital and Medical Center procured 10% green electricity and saved more than $190,000 per year, compared to its previous utility contracts.

Price discovery and consumer education for buyers. Reverse auctions allow buyers to quickly test the current market and easily compare prices between products before making a purchase decision. Buyers become more educated consumers as they learn about their electricity usage, product options, and prices as a result of participating in the auction process.

Reduced administrative burden. Participating in reverse auctions is faster and less burdensome for bidders than responding to RFPs. Buyers avoid lengthy proposal review processes.

Fast award and contract times. Contract terms and other details are set by buyers and announced to bidders ahead of auction day. Contract negotiations are limited and contracts can be signed quickly. Often, contracts between buyers and the winning bidder can be signed on the same day as the auction.

6.4. Challenges of Reverse Auctions

Due to the cost and logistics of implementing a reverse auction, this mechanism is generally available only to customers with high electricity demand or customers who are able to aggregate their demand with others, which limits its applicability.

Reverse auctions must be carefully designed in order to ensure fairness amongst bidders and in order to encourage the lowest possible price for customers. Thus, reverse auctions are often coordinated by third parties for a fee.

SUMMARY AND CONCLUSION

Increasingly, communities, businesses, non-profits, and utilities are partnering to provide a wider variety of options for procuring renewable energy. CCA programs, community wind and solar programs, green power community challenges, local collaborative electricity purchasing, bulk purchasing of on-site renewables, and reverse auctions provide unique methods, often resulting in cost savings and expanded access for consumers. The mechanisms are summarized below, as well as in Table 6.

Community choice aggregation programs, though not available in every state, allow communities to select an alternative energy supplier while continuing to receive transmission and distribution service from their existing supplier. While not all CCAs select renewable resources, programs such as MCE in California, Cape Light Compact in Massachusetts, and Oak Park in Illinois all offer renewable options.

Table 6. Overview of Innovative Renewable Energy Procurement Mechanisms

Innovative Mechanism	Sector Developing the Method	Procuring Sector	Benefits of Approach	Challenges to Approach
Community Choice Aggregation	Local governments	Utility consumers	• Potential lower cost • Large impact • Easy for consumers • Access to lower-cost, tax-exempt debt	• Need enabling legislation • Start-up costs can be significant • Possible resistance from incumbent electricity supplier • Need for education and outreach
Community wind and solar programs	Typically municipal or cooperative utilities and/or third-party project developers	Utility consumers, local residents, residents from anywhere	• Expanded access • Easy to sign up • Cost efficiencies of larger project or procurement • May offer a unique, local product • Can move subscription within service territory	• Tax and financial issues • Concerns over potential securities regulation
Green power challenges and local collaborative electricity procurement	Partnership between utility, local government, and/or large institutional purchasers	Utility consumers	• Contributes to branding of community • Provide education and outreach to community members • Easy to sign up • May encourage supplier competition • May lower cost	• Setting appropriate, realistic targets • Distributing costs among partner organizations • Administrative issues

Table 6. (Continued)

Innovative Mechanism	Sector Developing the Method	Procuring Sector	Benefits of Approach	Challenges to Approach
Bulk purchasing of distributed energy systems	Government agencies, businesses, non-profit organizations	Local residents, government agencies, schools, businesses	• Lower-cost solar • Reduced administrative effort for buyers • Improved company reputation for sponsors • Flexibility of program design and implementation • Quality assurance	• Requires a champion to coordinate • Potentially reduced flexibility for individuals in project design or contract terms
Reverse auctions	Businesses, governments, non-governmental organizations	Businesses, government agencies, non-profit organizations	• Increased price transparency and competition • Price discovery and buyer education • Reduced administrative burden • Fast award and contract times	• Pure reverse auction may not allow sufficient time to generate lowest bid • Relies on auction software technology

San Francisco is preparing to launch a renewable offer in 2012, and 242 communities in Illinois have voted to move forward with a CCA, though whether they will offer renewable energy is yet to be determined. CCAs can provide a lower-cost electricity product, large environmental impact, an easy way for consumers to support renewables, and access to lower-cost, tax-exempt debt.

CCAs can be challenging to implement because they require state legislation, can have significant start-up costs, face possible resistance from the incumbent electricity supplier, and because residents may be unaccustomed to thinking about their electricity use.

Community programs allow customers to invest collaboratively in renewable energy. In community solar programs, all participants purchase a share of a solar system and receive the financial benefits of the energy produced by their share. While few programs existed as late as 2008, by 2012, nearly 30 programs were identified. Community wind programs are emerging in the United States and are more common in Europe. Community programs expand access to renewable energy, provide an easy way to sign up, take advantage of cost efficiencies from larger projects or procurements, may offer a unique, local product, and may be more flexible, allowing participants to keep their investment, even if they move.

However, programs can be challenging to establish, due to tax, finance, and legal concerns; developing the program structure can be more challenging than finding subscribers.

Green power challenges and local collaborative electricity purchasing programs expand access to utility green pricing and competitive supplier programs. Local governments often partner with local utilities, third-party marketers, and/or environmental organizations to increase awareness of green pricing options through a green power challenge.

EPA's Green Power Communities are collectively buying more than 3.3 million MWh of green power annually, representing 9% of all voluntary sales in 2010. Groundswell's Community Power Project is a local collaborative electricity purchasing program that has facilitated procurement of 56,000 MWh of renewable energy for 109 organizations in the Washington, D.C., area.

Green power challenges and local collaborative electricity purchasing programs raise awareness and expand access to renewable options, provide alternative ways to sign up for renewable energy, may encourage friendly competition among suppliers, and can contribute to branding of a community. Challenges and procurement programs are fairly easy and straightforward to implement.

One barrier may be organizing the timing of the challenge and clarifying funding responsibilities between organizations. Local collaborative electricity procurement programs require a lead organization to educate potential partners and develop the actual procurement.

Group on-site purchasing programs identify a group of individuals or companies that are interested in installing solar systems and aggregate their demand for the equipment and installation services. This can result in attractive bulk pricing from vendors. In addition, the programs save buyers the time and effort that would be required to navigate the purchasing process individually, which may encourage more customers to install renewable energy systems.

The program may also include a financing aspect, which can further ease the process for customers and help expand the market. A group purchasing program requires a clear leader to organize the process. Successful programs have been organized and offered by a variety of leader organizations representing the private and public sectors, including the cities of Portland and San Jose, a citizen group in Los Angeles, a nation-wide enterprise called One Block Off the Grid, and the Orlando Utilities Commission.

Reverse auctions provide a flexible method for organizations and companies in restructured electricity environments to contract for electricity provision, including electricity from renewable energy sources. Customers in any electricity market can purchase RECs or contract for the construction of a renewable energy system using the reverse auction mechanism. Although many customers make use of paid auction organizers (e.g., World Energy Solutions), buyers with significant demand may choose to organize their own auctions, as the GSA has recently done. The transparency of the online bidding process often draws in a variety of potential bidders. The competitive environment created by reverse auctions has yielded significantly reduced prices for renewable energy, compared to traditional contracting processes. Careful pre-auction analysis and planning pays off in the increased confidence of buyers and bidders, as well as rapid post-auction contract signing.

8. Resources and References

Resources

Community Choice Aggregation

Illinois Community Choice Aggregation Network: http://www. ippconnect.com/CCA.php.

Local Energy Aggregation Network (LEAN Energy U.S.): www. leanenergyus.org.

Community Wind and Solar

Coughlin, J.; Grove, J.; Irvine, L.; Jacobs, J.F.; Johnson Phillips, J.; and Wiedman, J. (2012). *A Guide to Community Shared Solar: Utility, Private, and Non-Profit Project Development*. DOE/GO-102012-3569. Golden, CO: National Renewable Energy Laboratory. http://www.nrel. gov/docs/ fy12osti/54570.pdf.

Kubert, C. "Community Wind Financing." A Handbook by the Environmental Law & Policy Center. Accessed September 11, 2012: http://www.elpc.org/documents/WindHandbook2004.pdf.

Northwest SEED and Bonneville Environmental Foundation. (n.d.). "The Northwest Community Solar Guide." Accessed July 18, 2012: http://www. nwseed.org/documents/NW%20Community% 20Solar% 20Guide.pdf.

The Minnesota Project. (December 2009). "Lessons and Concepts for Advancing Community Wind." Accessed July 18, 2012: http://dev7. windustry. org/sites/dev7.windustry.org/files/Advancing-Community-Wind_Dec09.pdf.

Windustry offers case studies and links to guidebooks and other information regarding community wind project development on their website at www.windustry.org.

Green Power Challenge and Local Collaborative Electricity Procurement

EPA. "EPA Green Power Communities." Webinar. Accessed July 18, 2012: http://epa.gov/greenpower/events/23feb10_webinar.htm.

Groundswell: www.groundswell.org.

U.S. Environmental Protection Agency (EPA). "Green Power Communities." Accessed July 18, 2012: http://epa.gov/greenpower/ communities/ index. htm.

Bulk Purchasing of Distributed Energy Systems

DOE. (2011). "The Solarize Guidebook: A Community Guide to Collective Purchasing of Residential PV Systems." Accessed July 18, 2012: www.nrel.gov/docs/fy11osti/50440.pdf.

EPA. (2011). "Clean Energy Collaborative Procurement Initiative." Accessed July 18, 2012: http://www.epa.gov/greenpower/ cecp/ index.htm.

U.S. Department of Energy (DOE). (2011). "Solar Powering Your Community: A Guide for Local Governments." Accessed July 18, 2012: http://www4.eere.energy.gov/solar/sunshot/resource_center/sites/default/files/solar-powering-your-community-guide-for-local-governments.pdf.

World Resources Institute, Joint Venture, and Optony. (2011). Purchasing Power: Best Practices Guide to Collaborative Solar Procurement." Accessed July 18, 2012: http://www.wri.org/ publication/ purchasing-power.

Reverse Auctions

EPA. "Procuring Green Power Through Reverse Energy Auctions." Webinar. Accessed July 18, 2012: http://epa.gov/greenpower/events/ 24jun10_ webinar.htm.

Practice Greenhealth. "Healthcare Clean Energy Exchange." Accessed July 18, 2012: http://practicegreenhealth.org/topics/energy-water-and-climate/ energy/healthcare-clean-energy-exchange.

Practice Greenhealth. "Healthcare Renewable Energy Initiative." Accessed July 18, 2012: http://practicegreenhealth.org/topics/energy-water-and-climate/energy/healthcare-renewable-energy-initiative.

References

Bay Area Climate Collaborative (BACC). (2012). "SunShares." Accessed September 12, 2012: http://baclimate.org/impact/sunshares. html.

Bolinger, M. (2011). *Community Wind: Once Again Pushing the Envelope of Project Finance*. LBNL-4193E. Berkeley, CA: Lawrence Berkeley National Laboratory. http://eetd.lbl.gov/EA/EMP/reports/lbnl-4193e.pdf.

Braly, M. (9 May 2011). "*The Emerging Market for Small Renewables in California*." Accessed July 18, 2012: http://www.renewableenergy world.com/rea/news/article/2011/05/the-emerging-market-for-small-renewables-in-california.

California Public Utilities Commission (CPUC). (2010a). "*CPUC Puts PG&E on Notice Over Violations of Community Choice Rules*." Accessed April 18, 2012: http://docs.cpuc.ca.gov/word_pdf/NEWS_RELEASE/ 117229.pdf.

CPUC. (2010b). "*CPUC Refined Utility Marketing Rules for Community Choice Aggregation*." Accessed April 18, 2012: http://docs.cpuc.ca.gov/ PUBLISHED/NEWS_RELEASE/118215.htm.

EcoTech Centre. (2012). "*The Wind Turbine at the Ecotech Centre*." Accessed September 12, 2012: http://www.ecotech.org.uk/turbine.html.

Farrell, J. (2011). "*Group Purchase Gets Residential Solar to Grid Parity in Los Angeles*." Accessed May 8, 2012: http://cleantechnica. com/2011/ 12/05/group-purchase-gets-residential-solar-to-grid-parity-in-los-angeles/.

Fastman, B. (2011). "*CEC Launches Group Buy Solar Program*." Accessed May 8, 2012: http://www.independent.com/news/2011/apr/28/ cec-launches-group-buy-solar-program/.

Green Power Network. (2011a). "*New Community Solar Projects Launch in Colorado and Washington*." Accessed August 16, 2012: http://apps3.eere.energy.gov/greenpower/news/news_template.shtml?id=1 648.

Green Power Network. (2011b). "*Salt River Project Launches Community Solar Program*." Accessed August 16, 2012: http://apps3. eere.energy. gov/greenpower/news/news_template.shtml?id=1666.

Groundswell. (2012). "*Community Power Project*." Accessed June 1, 2012: http://groundswell.org/programs/community-power/about.

Heeter, J. & Bird, L. (2011). *Status and Trends in U.S. Compliance and Voluntary Renewable Energy Certificate Markets (2010 Data)*. NREL/TP-6A20-52925. Golden, CO: National Renewable Energy Laboratory. Accessed May 31, 2012: http://apps3.eere.energy.gov/ greenpower/ pdfs/52925.pdf.

Hinckely, T. (23 February 2010). "*Green Power Communities Webinar Transcript*." Accessed July 19, 2012: http://www.epa.gov/greenpower/ documents/events/gpc_webinar_transcript_022310.pdf.

Illinois Commerce Commission (ICC). (2012). "*Municipal Aggregation*." Accessed April 25, 2012: http://www.icc.illinois.gov/ ORMD/ MunicipalAggregation.aspx.

Joint Venture: Silicon Valley Network. (2011). "*SV-REP Documents and Resources*." Accessed July 19, 2012: http://www.jointventure.org/ index.php?option=com_content&view=article&id=524&Itemid=287.

Joyce, D. (2010). "*Procuring Green Power Through Reverse Energy Auctions*." U.S. EPA webinar. Accessed July 19, 2012: http://www.epa.gov/greenpower/events/24jun10_webinar.htm.

Lantz, E. & Tegen, S. (2009). *Economic Development Impacts of Community Wind Projects: A Review and Empirical Evaluation.* NREL/CP-500-45555. Golden, CO: National Renewable Energy Laboratory. http://www.nrel.gov/docs/fy09osti/45555.pdf.

LEAN Energy U.S. (2012). Accessed July 19, 2012: http://www.leanenergyus.org/what-is-cca/.

Lyderson, K. (2012). "Experts: Chicago Aggregation Could Hurt Renewable Energy – Unless RPS is Fixed." *Midwest Energy News*. Accessed August 16, 2012: http://www.midwestenergynews.com/2012/ 08/10/experts-chicago-aggregation-could-hurt-renewable-energy-unless-the-rps-is-fixed/.

MadiSUN Commercial Solar Group Buy (MadiSUN). (2012). Accessed September 12, 2012: http://madisungroupcom.wordpress.com/.

Marin Clean Energy (MCE). (2008). *Blueprint for the Future: A Citizens' Guide to the Proposed Marin Clean Energy Authority*. Accessed July 19, 2012: http://www.marincommunityenergy.org/files/ DRAFT%20Marin%20Clean%20Energy%20Citizens'%20Guide%20042408v5%20copy.pdf.

Marin Clean Energy (MCE). (2012). "*Our Story*." Accessed January 31, 2012: http://www.marincleanenergy.com/index.php/our-story/our-team. html.

Marin Energy Authority (MEA). (2011). "Revised Community Choice Aggregation Implementation Plan and Statement of Intent." Accessed February 9, 2012: http://www.marincleanenergy.com/ images/stories/PDF/Implementation_Plan_Revised_12.22.11.pdf.

Marshall, S. (2012). "*Greening the Grid Through Community Choice Aggregation*." Accessed March 12, 2012: http://epa.gov/greenpower/documents/events/6mar12_marshall_presentation.pdf.

Midwest Renewable Energy Association. (2011). "*Wisconsin Organizations Awarded DOE SunShot Grant*." Accessed May 8, 2012: http://ci.marshfield. wi.us/files/SunShotGrant.pdf.

Natural Marketing Institute. (April 2011). *Consumer Attitudes About Renewable Energy: Trends and Regional Differences*. NREL/SR-6A20-50988. Golden, CO: National Renewable Energy Laboratory. Accessed October 27, 2011: http://apps3.eere.energy.gov/greenpower/pdfs/50988.pdf.

New Centennial Power. (2012). "*Huerfano River Wind Project*." Accessed June 28, 2012: http://www.newcentennialpower.com/.

Orlando Utilities Commission (OUC). (2011). "*Pilot Solar Aggregation Project.*" Accessed September 12, 2012: http://www.ouc.com/Libraries/Supplier_Documents/RFP3030-FINAL.sflb.ashx.

Pizzi, R. (2009). "Online Energy Auctions Save Money for Chicago Hospital." *Healthcare Finance News.* Accessed May 8, 2012: http://www.healthcarefinancenews.com/news/online-energy-auctions-save-money-chicago-hospital.

Practice Greenhealth. (2012). "*Healthcare Clean Energy Exchange (HCEE).*" Accessed May 8, 2012: http://practicegreenhealth.org/topics/ energy-water-and-climate/energy/healthcare-clean-energy-exchange.

Revkin, A. (8 March 2012). Postcard From a Solar Rooftop. *New York Times.* Accessed August 15, 2012: http://dotearth.blogs.nytimes.com/2012/03/08/postcard-from-a-solar-rooftop/.

Sabatini, J. (2 October 2011b). "SF Green Public Power Launch Gets Price Tag: $19.5 Million." The Examiner. http://www.sfexaminer.com/local/2011/10/sf-green-public-power-launch-gets-price-tag-195-million.

Sabatini, J. (24 November 2011a). "CleanPowerSF Will Still Rely on Fossil Fuels Despite Claims." *The Examiner.* Accessed September 11, 2012: http://www.sfexaminer.com/local/2011/11/cleanpowersf-will-still-rely-fossil-fuels-dispite-claims.

Shah, C. (2011). "Renewable Power Purchases and Renewable Energy Certificates (RECs)." Federal Energy Management Program, U.S. Department of Energy. Accessed May 8, 2012: http://www1.eere.energy.gov/femp/pdfs/rec_webinar_062311.pdf.

Sherwood, L. (2012). "U.S. Solar Market Trends 2011." *Interstate Renewable Energy Council.* Accessed August 15, 2012: http://www. irecusa.org/wp-content/uploads/IRECSolarMarketTrends-2012-web.pdf.

Simes, R. (2012). "*Cincinnati Exploring 100% Renewable Energy Plan.*" Accessed July 19, 2012: http://www.bizjournals.com/cincinnati/ blog/2012/02/cincinnati-exploring-100-renewable.html.

Soares, J. (5 March 2012). Email. Cape Light Compact, Barnstable, MA.

Speer, B. (2011). "The New(er) Kids on the Block: Community Choice Aggregators." Accessed August 15, 2012: https://financere.nrel.gov/finance/content/newer-kids-block-community-choice-aggregators.

The Minnesota Project. (2009). "Lessons & Concepts for Advancing Community Wind." Accessed June 28, 2012: http://dev7.windustry.org/sites/dev7.windustry.org/files/Advancing-Community-Wind_Dec09.pdf.

U.S. Department of Energy. (2011a). "The Solarize Guidebook." Accessed July 19, 2012: http://www.portlandonline.com/bps/index.cfm? &c=54114.

U.S. Department of Energy. (2011b). "Solar Powering Your Community: A Guide for Local Governments." Accessed July 19, 2012: http:// solaramericacommunities.energy.gov/pdfs/Solar-Powering-Your-Community-Guide-For-Local-Governments.pdf.

U.S. Environmental Protection Agency (U.S. EPA). (2010). "Improving Solar PV Results Through Collaborative Procurement." Webinar. Accessed July 19, 2012: http://www.epa.gov/greenpower/events/ 4aug10_webinar.htm.

U.S. EPA. (2011). "Clean Energy Collaborative Procurement Initiative." Accessed July 19, 2012: http://www.epa.gov/greenpower/cecp/ washington.htm.

U.S. EPA. (2012a). "Green Power Partnership. Purchase Requirements." Accessed September 6, 2012: www.epa.gov/greenpoer/join/purchase. htm.

U.S. EPA. (2012b). "Green Power Partnership. Green Power Community Challenge Rankings." Accessed September 6, 2012: www. epa.gov/greenpower/communities/greenpower/communities/gpcrankings.h tm.

Windustry. (2012). "The Minnesota Flip." Accessed September 11, 2012: http://www.windustry.org/community-wind/toolbox/chapter-12-minnesota-flip.

Witherbee, S. (5 June 2012). Email. Groundswell. Washington, D.C.

World Energy. (2012). "GSA Saves New York Federal Facilities Over $35 Million in Power Costs Through World Energy Solutions." Accessed May 8, 2012: http://www.worldenergy.com/news/gsa-saves-new-york-federal-facilities-over-35-million-in-power-costs-through-world-energy-solutions/.

9. APPENDIX

Table A-1. Overview of Community Solar Programs

Utility/Provider	Program Name	Program Size (kW)	Program Start Year
Ellensburg (WA)	Community Solar Project	111	2006
Sacramento Municipal Utility District (CA)	SolarShares	1,000	2008
Florida Keys Electric Cooperative (FL)	Simple Solar	117	2008
Ashland (OR)	Solar Pioneers II	64	2008

St. George (UT)	SunSmart	250	2009
Bainbridge Island (WA)	Solar for Sakai	5	2009
Holy Cross Energy/Clean Energy Collective (CO)	Mid Valley Solar Array (El Jebel)	80	2010
University Park Community Solar LLC (MD)	University Park Solar	23	2010
Corvallis (OR)	Corvallis OR, SunSlice Deal	2	2010
Okanogan County Electric Cooperative (WA)	OCEC Community Solar	20	2010
Salt River Project (AZ)	SRP EarthWise & Copper Crossing Solar Ranch	2,000	2011
Trico Electric (AZ)	Trico SunWatts Sun Farm	193	2011
Delta-Montrose Electric Association (CO)	Community Solar Array	20	2011
Holy Cross Energy/Clean Energy Collective (CO)	Garfield County Array	858	2011
Berea Utilities (KY)	Berea Solar Farm	14	2011
Edmonds (WA)	Edmonds Community Solar Cooperative	4	2011
Okanogan County Electric Cooperative (WA)	Winthrop Community Solar	23	2011
Poulsbo Project (WA)	Poulsbo Middle School	75	2011
Seattle City Light (WA)	Community Solar	24	2011
Tuscon Electric Power (AZ)	Bright Tucson Community Solar Program	1,600	2011
UniSource Energy Services (AZ)	Bright Arizona Community Solar	1,720	2012
Colorado Springs (CO)	Community Solar Garden Facility Incentive Program	500	2012
Poudre Valley REA/Clean Energy Collective (CO)	Community Solar	115	2012
Brewster Community Solar Garden Cooperative Inc. (MA)	Brewster Community Solar Garden	346	2012
Olympia (WA)	Olympia WA, SunSlice Deal	75	2012
Acorn Energy Cooperative (VT)	Acorn Energy Solar One	150	2012
United Power (CO)	Sol Partners Cooperative Solar Farm	10	2012
San Miguel Power Association/Clean Energy Collective (CO)	SMPA Community Solar	1,000	2012

End Notes

[1] The guide was produced by a collaborative effort between the U.S. EPA, the U.S. Department of Energy, the World Resources Institute, and the Center for Resource Solutions. http://www1.eere.energy.gov/femp/pdfs/purchase_green_power.pdf.

[2] Though not discussed in this paper, CCAs exist in Rhode Island and Ohio, but they do not offer substantial renewable energy products. The Northeast Ohio Public Energy Council (NOPEC) aggregates natural gas and electricity consumer load, and the Rhode Island Energy Aggregation Program (RIEAP) aggregates load of local governments in the state. RIEAP renewable percentages have ranged from 5% to 10% (LEAN Energy U.S. 2012).

[3] For example, in Illinois, opt-out programs can be implemented but must be approved through a referendum question at a general election.

[4] For more on securities issues, see Coughlin et al 2012.

[5] The Bay Area Climate Collaborative is a public-private initiative developed by the Silicon Valley Leadership Group to encourage a clean-energy economy. See http://www.baclimate.org.

[6] The Public Sector Climate Task Force is made up of sustainability officers from across the region.

[7] While the focus here is the use of reverse auctions in voluntary renewable energy markets, the mechanism is also being used in compliance markets. For example, the State of Connecticut's Public Utilities Regulatory Authority approved the ZREC (zero-emissions) and LREC (low-emission) renewable energy program. Through the program, state utility companies will hold a series of reverse auctions that will result in long-term contracts to purchase the RECs associated with electricity generated by customers of the state's distribution companies. The program was mandated by the state Energy Act of 2011. See http://www.ct.gov/pura/cwp/view.asp?A=4144&Q=502078 and http://energy.aol.com/2012/04/30/connecticut-focuses-on-unique-reverse-auctions-to-drive-green/.

In: Voluntary Markets in Renewable Energy ISBN: 978-1-62417-561-9
Editor: Jamir S. Fuller

Chapter 3

VOLUNTARY GREEN POWER MARKET FORECAST THROUGH 2015*

Lori Bird, Ed Holt, Jenny Sumner and Claire Kreycik

EXECUTIVE SUMMARY

Voluntary markets for renewable energy in which consumers purchase renewable energy from their electricity providers or from renewable energy certificate (REC) marketers have existed for more than a decade. Going forward, various factors will influence the development of these markets, including potential climate policies, state and federal renewable portfolio standards (RPS), renewable energy prices, the level of consumer interest in purchasing green power, and the interest on the part of utilities in offering new green options and in continuing to promote existing programs.

This report presents estimates of voluntary market demand for renewable energy or green power through 2015. The forecast relies on historical data, including market sector sizes and growth rates, to inform our assumptions. Though we adjusted growth rates as discussed in the report, they remain the starting point for our analysis. Because of the different factors that affect utility programs, REC markets, and competitive electric markets, separate assumptions and methods are used to project demand for each submarket.

---

* This is an edited, reformatted and augmented version of the National Renewable Energy Laboratory, NREL/TP-6A2-48158, dated May 2010.

We developed a low-growth and a high-growth scenario for each market segment. The low-growth scenario reflects factors that may limit growth over time. The high-growth scenario represents optimistic growth assumptions. The difference between the scenarios reflects market and policy uncertainty and the difficulty predicting how consumers will react to factors affecting the voluntary market. The analysis includes a negative policy impacts scenario designed to reflect impacts on the market that might occur if future carbon policies significantly reduce the motivation for purchases. This scenario reflects the potential loss of a significant volume of existing purchasers.

Given the assumptions used in each market sector, our forecast projects the total voluntary demand for renewable energy in 2015 to range from 63 million MWh annually in the low-growth scenario to 157 million MWh annually in the high-growth scenario (Figure E-1). The range represents an approximately 2.5-fold difference. The negative policy impacts scenario reflects a market size of 24 million MWh. When extended through 2020, projections show 327 million MWh annually in the high-growth scenario and 94 million MWh annually in the low-growth scenario. This range is greater than the 2015 range, representing an approximately 3.5-fold difference. The negative policy impacts scenario projects 33 million MWh in 2020.

In both the high-growth and low-growth forecasts, REC markets continue to drive much of the growth, though the contribution of RECs is significantly less in the negative policy impacts scenario. On-site solar is projected to increase its share of the total voluntary market, while utility green power programs and competitive markets are projected to decline as a share of the total voluntary market.

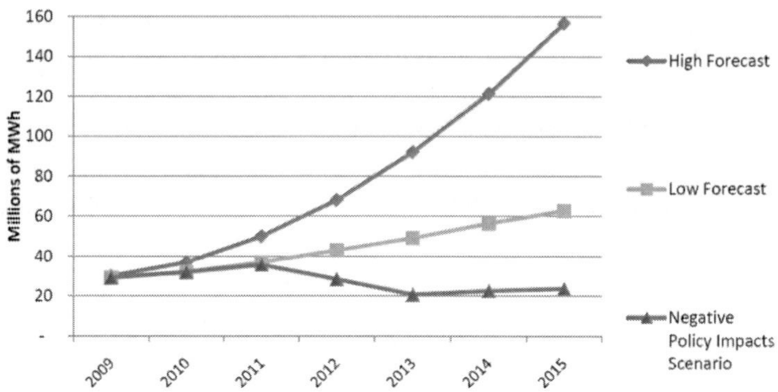

Figure E-1. Total voluntary demand for renewable energy (millions of MWh): High-growth, low-growth, and negative policy impacts scenario forecasts through 2015.

Several key uncertainties affect the results of this forecast, including uncertainties related to growth assumptions, the potential impacts of policies on markets, the price and competitiveness of renewable generation, and the level of interest that utilities have in offering and promoting green power products.

1. INTRODUCTION

Voluntary markets for renewable energy in which consumers purchase renewable energy for their electricity needs from their utilities or electricity providers have existed since the mid 1990's. Residential consumers participate in these markets and buy renewable energy for a variety of reasons: environmental concerns, including global climate change; energy security and independence; support for clean energy technologies; and conservation of resources for future generations. Businesses may have the same concerns as residential consumers but may also be motivated to differentiate their products, meet corporate environmental goals, or satisfy stakeholders.

Any consumer can purchase "green power"[1] through renewable energy certificates (RECs). RECs represent the "environmental attributes" of electricity generated from renewable energy-based projects. "Unbundled RECs" refer to RECs that are sold separately from electricity generation, thus allowing a consumer to support renewable energy without switching to an alternative electricity supplier. REC market sales reflect only those sales to end-use consumers that are separate from electricity.

The vast majority of RECs are sold to commercial and institutional customers, which have increasingly been driving the market. The Environmental Protection Agency's (EPA's) Green Power Partnership, a voluntary program that encourages organizations to purchase green power, has seen significant growth in recent years. The Partnership has grown from 104 partners in May 2003 to 1,203 partners in October 2009. Organizations report the quantity of RECs they purchase, which has increased from approximately 83,400 megawatt-hours (MWh) in May 2003 to approximately 13.3 million MWh in December 2009 (EPA 2010).

In addition to RECs, consumers and businesses in regulated utility markets may have the option to purchase some portion of their power supply as renewable energy. Regulated utility programs, or "green pricing programs" are offered by more than 850 utilities, or about 25% of utilities nationally. Finally, in competitive (or restructured) retail electricity markets, consumers may have the opportunity to purchase electricity generated from renewable sources by switching to an alternative electricity supplier that offers green power.

The total voluntary market for green power exceeded 24 million MWh in 2008. From 2007 to 2008, voluntary green power sales increased by 34% and annual growth rates have averaged 41% since 2004. From 2007 to 2008, REC markets grew by 47%, utility green pricing programs grew by 12%, and competitive markets grew by 22% (Bird et al. 2009a).

A variety of factors will influence the development of these markets going forward, including potential climate policies, the development and implementation of state and federal renewable portfolio standards (RPS) and renewable energy price impacts, the level of consumer interest in purchasing green power, and interest on the part of utilities in offering new green options and in continuing to promote existing programs. A previous forecast of the voluntary market (Wiser et al. 2001) projected that green power demand in 2010 would range from 905 average megawatts (7.9 million MWh) in a low-growth scenario to a 6,971 average megawatts (61 million MWh) in a high-growth scenario. This analysis presents an updated forecast based on market experience from the last decade and addresses new market influences and trends.

This report presents estimates of voluntary market demand for renewable energy through 2015. Because of the different factors that affect utility programs, REC markets, and competitive electric markets, separate assumptions and methods are used to project demand for each submarket. A low-growth and a high-growth scenario were developed for each market segment, with the low case reflecting factors that may limit growth over time and the high-growth scenario representing optimistic growth assumptions. The analysis also includes a scenario in which future carbon policies have significant negative impacts on voluntary market demand.

2. METHODOLOGY

In forecasting demand for green power, we developed a spreadsheet model to estimate future market size based primarily on historical market trend data and factors that will influence uptake rates and offerings in the near term. This approach was used because of the availability of detailed historical market data on participation rates, purchase volumes, and product offerings, which provide a reasonable basis for forecasting trends over the near term, which is the focus of the study. It is not a mechanistic approach, however, in that we used judgment in adapting the historical data for our growth assumptions based on market maturity and potential policy impacts.

In the analysis, we treated each of the three market segments separately: unbundled RECs, regulated utility programs, and competitive electricity markets. Current market shares for each segment are shown in Figure 1. Sales from REC markets (15.6 million MWh) represented nearly two-thirds of voluntary demand in 2008. Utility green pricing programs and competitive markets represented 20% (4.8 million MWh) and 16% (3.9 million MWh) of the 2008 market, respectively (Bird et al. 2009a).

For each of these submarkets, we used separate assumptions based on market differences and the historical data available for each, which are described in detail below. For each of these submarkets, we developed two scenarios: 1) a high-growth scenario with optimistic assumptions about offerings and participation levels, and 2) a low-growth scenario that assumed modest growth estimates to reflect potential policy impacts as well as more limited interest among utilities and retail suppliers in offering and promoting options and among consumers in purchasing green power. We also examined a negative policy- impacts scenario in which carbon policy design has significant negative impacts on the market.

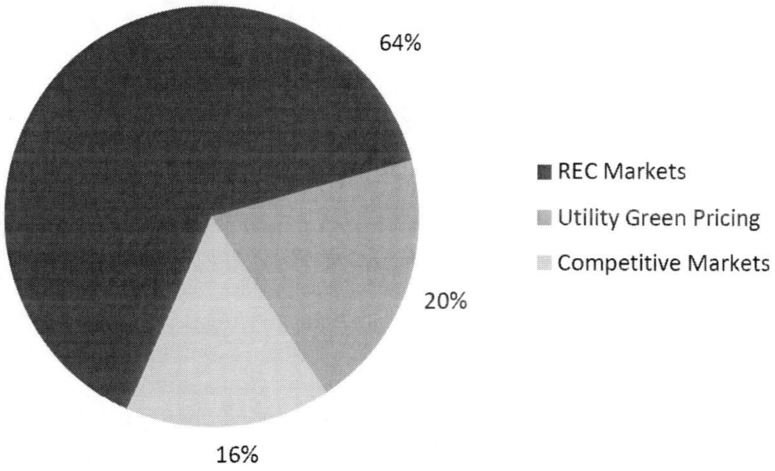

Source: Bird et al. (2009a).

Figure 1. Share of 2008 green power sales by market segment.

The focus of this work is a projection of voluntary demand through 2015, though a projection was also made through 2020. While the 2020 results are presented in the summary (Section 8), there is greater uncertainty in these estimates, particularly because this forecast relies heavily on historical growth rates. There is also significant policy uncertainty over the timeframe that can impact market growth; therefore, we focus on the 2015 estimates.

3. Scenario Considerations and Assumptions

A number of external factors may affect growth in voluntary demand for renewable energy in the future, some by depressing demand in the low-growth scenario, and some by increasing demand in the high-growth scenario. The three market demand scenarios examined in this analysis are described below and summarized in Table 1.

3.1. High-Growth Scenario

The high-growth scenario (high case) reflects optimistic assumptions about growth in green power purchasing in coming years. It assumes that growth continues at fairly rapid rates comparable to those in recent years and that any potential federal or state policies will have no significant negative impacts on these markets. The high case assumes that there will be sufficient renewable energy supplies to meet demand and that any barriers to development are tractable so that voluntary market demand may be uninhibited.[2] Under this scenario, REC markets continue to grow strongly as prices remain relatively low and motivation for purchasing remains strong. This scenario also assumes utilities undertake renewed efforts to market green power and to offer new products to consumers, which could be spurred in part because utilities find that voluntary programs help them achieve portfolio diversity and reduce risks of future compliance through the purchase or ownership of renewables. Further, utilities may find that addressing both RPS and voluntary programs offers them a risk management strategy whereby they can acquire renewable resources in larger quantities, and in advance of compliance needs at lower cost. The high case also assumes optimistic growth in on-site renewable energy installations driven by aggressive cost reductions in solar photovoltaics. Finally, this case assumes that cap and trade legislation that supports the ability of renewable energy to reduce greenhouse gas (GHG) emissions is adopted, and this scenario specifically supports environmental claims by purchasers of renewable energy.[3] Because this would remove current uncertainty, it could lead to greater enthusiasm for purchasing renewable energy.

3.2. Low-Growth Scenario

The low-growth scenario (low case) reflects modest assumptions about growth in green power purchasing in coming years. It assumes that green power markets grow significantly more slowly than in the past because of reduced interest in purchasing and promoting products and because of potential price increases. It assumes that growing renewable energy mandates, at either the state or federal level, increase competition for limited renewable resources, leading to higher REC prices, which slow demand growth. Demand may also slow if consumers feel that RPS or other renewable energy policies are adequately driving renewable energy generation. The low case also assumes that utilities are less inclined to offer and promote voluntary green power programs. This might occur if utilities struggle to meet RPS targets or if they do not see the need to promote the sale of renewable energy as a differentiated product and instead decide simply to include renewable energy acquisitions in rates for all customers. The low case assumes slower growth in on-site renewable energy installations, based on a scenario of more modest cost reductions in solar photovoltaics than in the high case. The low case assumes that even if carbon policy is adopted that purchasers are still able to use renewable energy purchases to address their electricity emissions in GHG inventory programs and to meet corporate environmental goals.

3.3. Negative Policy Impacts Scenario

The negative policy impacts scenario reflects a case in which future policies reduce motivation for purchasing green power and significantly reduce demand in coming years. In this scenario, the growth assumptions are generally the same as in the low case until policies are assumed to affect the market. This scenario reflects impacts on the market that could occur if cap and trade legislation is adopted that does not support the ability of renewable energy to reduce GHG emissions.

Table 1. Summary of Scenario Assumptions

High Case	Low Case	Negative Policy Impacts Case
Demand continues to grow rapidly and any price impacts from RPS are limited	RPS requirements are increased, creating competition for RECs and raising prices	RPS requirements are increased, creating competition for RECs
Potential carbon policies are designed to support emissions reductions claims by voluntary renewable energy purchasers (e.g., through allowance retirement)	GHG inventories allow green power purchases to count toward electricity emissions	Carbon policy is adopted that does not allow emission reduction claims and purchasers are not able to count green power purchases toward GHG emissions goals, causing many existing customers to cease purchasing.
Utilities actively promote programs	Utilities less inclined to promote programs	No new utility programs

The negative policy impacts scenario assumes that future carbon policy design does not support environmental claims by purchasers of renewable energy, which significantly reduces motivation to purchase, even among existing purchasers. If renewable energy purchasers cannot account for the GHG benefits of renewable energy purchases in GHG inventories or cannot make carbon-neutral or emissions benefits claims, organizations motivated by an interest in the carbon benefits of renewable energy would likely lose motivation to purchase (see Bird et al. (2007) and Harmon and Hirschhorn (2006) for additional discussion of this issue).[4] This scenario assumes customers interested in the carbon benefits of green power cease purchasing or do not commit to purchases if green power can no longer be used as a carbon mitigation strategy. Also, marketers may cease to offer products without carbon benefits or may feel it is deceptive to offer products in which consumers may believe those benefits exist.

Our forecast assumptions are influenced by these external factors but generally are not quantitatively tied to them. Further, our assumptions were influenced by an awareness of the typical life cycle of demand for new products. As voluntary markets mature, they can be expected to reflect the diffusion of innovation, or a standard "S" curve, starting slowly, and proceeding through rapid growth before tapering off (Holt and Wiser 1999). The challenge is that we do not know just where we are in this S curve, whether markets will experience considerably more growth in demand before growth flattens, or whether demand is nearing the top of the S curve and will soon turn flat. The high case would reflect a position in the steeper part of the curve, while the low case would reflect a point nearer the top.

4. UNBUNDLED REC MARKETS

This section describes the methodology and assumptions used to estimate the size of REC markets in coming years. Subsequent sections of the report discuss the methods used to estimate utility programs, competitive markets, and demand for on-site systems.

Voluntary demand for RECs is driven primarily by corporations and large organizations, with residential REC purchases accounting for only about 1% of REC sales (Bird et al. 2009a). We assumed future growth in REC markets would continue to be dominated by nonresidential consumers because of the cost and challenges of explaining and marketing RECs to residential consumers, who unlike more-sophisticated nonresidential buyers are generally unfamiliar with the functioning of electricity markets. Because RECs are often sourced from a national market and because data on regional demand are limited, we did not attempt to differentiate REC market demand by region.

In our forecast, we rely on historical nonresidential demand data from the Green Power Partnership to determine historical growth rates. U.S. Environmental Protection Agency (EPA) data are divided by size of customer load into four categories. The largest customers, by size of load, account for 83% of Partnership REC demand in 2009, as shown in Table 2, and these shares have been consistent with no more than 1% variation since 2006.

Table 3 shows growth rates for nonresidential customers in these load size categories and how they vary for different periods. Using rates prior to 2006 in the analysis would yield growth rates higher than are likely sustainable going forward. Some of the largest growth occurred in the early years of the Green Power Partnership program, and growth since then, though strong, has moderated somewhat. Table 3 shows somewhat slower growth from 2008 to 2009, presumably reflecting the economic recession.

Differences in the growth rates for the various customer size categories are generally not significant enough to affect overall growth rates substantially. This can be seen from the fact that the growth rates for the largest customer category are nearly the same as for the total customers in the Partnership. The growth rate for the smallest purchasers is considerably lower than for the other categories, but because they represent less than one percent of demand, using a different growth rate would have minimal effect on the outcome.

Table 2. 2009 REC Sales to Green Power Partners, by Size of Customer

Total Load of Customer	Percent of RECs Purchased	RECs Purchased (MWh)
>100,000 MWh	83%	11,164,620
10,001-100,000 MWh	14%	1,843,608
1,001-10,000 MWh	3%	437,755
<1,000 MWh	<1%	50,823

Source: EPA (2009), as of October 15.

Table 3. Average Annual Growth Rates for REC Sales

Customer Size (load)	2003*-2009	2006-2009	2007-2009	2008-2009
> 100,000 MWh	137%	53%	22%	9%
10,001 - 100,000 MWh	75%	51%	23%	3%
1,001 - 10,000 MWh	91%	65%	53%	35%
< 1,000 MWh	58%	36%	8%	6%
Total	114%	53%	23%	9%

Source: EPA (2010).

* Reflects year-end 2003 estimate based on May 2003 and July 2004 data.

Because EPA's Green Power Partnership data offer greater detail, we analyzed them for trends. However, our forecast used National Renewable Energy Laboratory (NREL) data on REC markets reported in Bird et al. (2009a) to estimate the total market size in 2009, which serves as the basis of the forward projections.[5]

High Case

For our high-growth scenario (or high case), we assumed 25% growth in 2010, consistent with growth from 2007 to 2009, to reflect the likelihood that the economic downturn will continue to inhibit demand in the near term. Some companies have reduced the size of their purchases during this period and others may be waiting for a better economic climate to make substantial size purchases. However, we project that demand will increase in 2011 based on the expectation of an improved economy.

For the period 2011-2013, we assumed 40% growth per year, which is generally consistent with pre-recession growth rates, but somewhat below average growth from 2006 through 2009 (55%). Because of the difficulty in sustaining continually high and compounding growth rates year after year, and to account for other factors that may slow growth, we assumed growth rates of 30% in 2014-2015, 20% in 2016-2017, and 10% in 2018-2020. Such factors include the possible shift of some REC demand to on-site systems and the assumption that the innovators and early adopters of the classic market diffusion curve have already chosen to participate, making it more difficult to attract new purchasers over time. However, the high case reflected generally modest impacts because of these factors over the period, with the low case reflecting greater impacts.

Low Case

For our low case assumption, we assumed demand for RECs would grow at a rate of 10% in 2010 to account for the possibility of greater impacts of the economic downturn on demand than in the high case. For the period 2011-2013, we assumed 15% growth, which is consistent with the slow growth rate experienced in 2009 and substantially lower than the historical 55% growth rate for the 3-year period, which includes pre-recession growth. Consistent with declining growth in the high case, we assumed growth rates would be 10% in 2014-2017, and 5% in 2018-2020.

The lower growth assumptions reflect, in part, the possibility that REC prices will rise because of competition with the growing compliance markets, which could dampen demand in the voluntary REC market. Lower growth assumptions also reflect the possibility that the market is more mature and attracting new purchasers will become more difficult. These assumptions translate into slower growth but do not affect the level of current purchases.

Negative Policy Impacts Scenario

Because REC markets have been driven by a number of corporate or institutional purchasers that use RECs for carbon accounting purposes or for achieving carbon-neutrality goals, the negative policy impacts scenario examines a case in which carbon policy design significantly decreases REC purchaser motivation. If carbon policy and GHG accounting rules are designed so that renewable energy purchasers cannot account for the GHG benefits of renewables in GHG inventories or make carbon-neutral or emissions benefits claims, organizations motivated by an interest in the carbon benefits of renewables would likely lose motivation to purchase. This could lead to current purchasers, particularly many of the largest green power purchasers, moving away from green power purchases as a carbon mitigation strategy. Because large purchasers dominate the REC market segment, and because they are expected to be most affected by policy design considerations as they are most likely to be influenced by carbon accounting rules and to use public carbon emission inventories, REC markets could be substantially impacted.

To our knowledge, no comprehensive or statistically significant data are available to indicate the fraction of purchasers motivated primarily or exclusively for the carbon benefits of renewable energy.[6] The data that are available, however, suggest that the ability to make carbon claims is a strong motivator for green power purchases by large organizations—those that are driving the growth in this market.

The Center for Resource Solutions looked at the public claims made by the largest 15 corporate purchasers in the EPA's Green Power Partnership and found that two-thirds of them made statements relating to reducing their emissions or carbon footprint (Busch 2010). Still, some organizations have indicated that they are primarily interested in making claims about products made with renewable energy, and others have suggested that they are likely to move toward greater on-site renewables over time. Because of the variety of possible motivations and rationales for purchasing,[7] responses to future carbon policy changes may also be mixed. Therefore, it is difficult to determine how much of an impact this will have on participation.

In the negative policy impacts scenario, we assumed growth would be similar to that in the low case in 2010 and 2011, prior to the implementation of carbon policy, with purchases increasing by 10% in each year. We assumed that starting in 2012—when proposed carbon policies are slated to take effect—and over the two-year period through 2013, that two-thirds of REC purchasers will cease to purchase RECs if they can no longer be used for carbon-neutrality claims or to address emissions in GHG emission inventories. Thus, two-thirds of the volume of REC purchases is phased out evenly over the two-year period. We assumed that in 2014 and onward the market will continue to decline at a rate of -10% annually.

Summary

Based on these assumptions, in 2015, the unbundled REC market is projected to be between 33.7 million MWh in the low case and 96.5 million MWh in the high case (Table 4).

The 2015 high case projection is 2.9-fold higher than the projection in the low case and represents a 5.8-fold increase from the 2009 REC market. In 2015, the negative policy impacts scenario represents a 65% decrease from 2008, resulting in 5.4 million MWh.

5. Regulated Utility Programs

To estimate growth in regulated utility green pricing programs, we relied on NREL data collected annually from utility green pricing programs. Data for a five-year period from 2004-2008 inclusive were used. The data were sorted by utility categories (investor-owned utilities (IOUs), public utilities (municipal utilities and public utility districts), and electric cooperatives), as well as by programs that worked with a marketing partner or not.

We differentiated our forecast assumptions for 1) existing utility programs and 2) new utility programs that may be added over the forecast period.

5.1. Existing Utility Programs

As shown in Table 5, the compound annual growth rates (CAGR) for the period 2004-2008 are not substantially different between IOUs and public utilities, although IOU programs grew at a slightly higher annual rate than public utility programs (29.2% vs. 24.5%). On average, cooperative programs did not grow during the period perhaps reflecting natural customer attrition and relatively lower levels of marketing once the programs are launched.[8]

Although not shown in the table, we also examined growth in programs that were supported by a third party marketer. An earlier study (Bird and Kaiser 2007) found that some programs (both IOU and public utilities) had larger customer participation if they worked with a marketing partner. However, as a group, growth in sales for utilities working with a marketer could not be differentiated from growth of those working without a marketer. For 2004-2008, utilities that partnered with marketers showed higher growth (29.4%) than those that did not work with marketers (26.3%), but for 2005-2008, the results are reversed (21.9% vs. 25.4%). As a result, we did not differentiate growth assumptions based on whether a marketing partnership is present.

High Case

For our high-case projection, we assumed IOUs would grow at an annual rate of 25% through 2014, with the exception of the 2009-2010 timeframe, where due to the downturn in the economy, we assumed growth of 10%.

Table 4. Historical and Projected Voluntary Demand for RECs (thousand MWh)

	2006	2007	2008	2009	2010	2011	2012	2013	2014	2015
High Case	6,800	10,600	15,600	16,700	20,800	29,100	40,800	57,100	74,300	96,500
Low Case	6,800	10,600	15,600	16,700	18,300	21,100	24,200	27,900	30,600	33,700
Negative- Policy Impacts Scenario	6,800	10,600	15,600	16,700	18,300	20,200	13,400	6,700	6,000	5,400

Table 5. Green Power Sales Growth Rates for Existing Programs, by Type of Utility

Year	IOUs		Public Utilities		Cooperatives*	
	Sector GP Sales (MWh)	Growth from previous year	Sector GP Sales (MWh)	Growth from previous year	Sector GP Sales (MWh)	Growth from previous year
2008	2,779,830	9.6%	2,019,864	20.8%	56,656	-7.4%
2007	2,535,199	10.9%	1,672,101	15.9%	61,198	-3.2%
2006	1,851,335	66.3%	1,443,051	37.4%	63,210	-19.3%
2005	1,374,616	37.7%	1,050,140	24.9%	78,364	33.6%
2004	998,491	-	840,701	-	58,649	-
CAGR (2004-2008)		29.2%		24.5%		-0.9%
CAGR (2005-2008)		26.5%		24.4%		-10.2%

*Adjusted for outlier, see footnote 7.

Although 25% growth is slightly less than the average actually experienced over a five-year period, we note that year-to-year growth from 2006-2007 and 2007-2008 was substantially lower than the earlier years, which may signal a slowing of growth not influenced by the later recession (although the 2007-2008 data may include some of the economic slowdown that began in 2008). To account for difficulties in attracting new customers over time and in sustaining high-growth rates over many years, we assumed growth slows to 20% in 2015-2017 and to 15% in 2018-2020.

For public utility programs, growth rates have been more consistent historically with no similar significant drop in growth in recent years. For this reason, we assumed a high-case growth of 25% through 2014 the same as for IOUs, even though the five-year average growth is slightly lower than that for IOU programs. We did not assume a slowdown in 2009-2010 because historical data do not show significant slowing from the economic downturn. Consistent with the IOU growth rates, we assumed growth slows to 20% in 2015-2017 and to 15% in 2018-2020 to account for difficulties in attracting new customers over time and in sustaining high-growth rates over many years.

Cooperative programs, based on our adjusted data, have seen slightly negative growth. Recent reductions in sales are within the bounds of typical program attrition as customers choose to leave the program or simply move away. For the 2009-2010 timeframe, we assumed growth would be consistent with 2008 observed trends, and in the high and low cases, we saw declines of 7%. For the high case, we assumed cooperative programs would grow by 10% per year from 2011 to 2014, and then decline to 5% annually from 2015 through 2020. Such growth is plausible if cooperative members show interest in green power and their management supports increased promotion and attention to the program. The reduction in 2015 is designed to account for the challenge in continuing to attract new customers as the market matures.

Low Case

The low case growth rate for existing IOU programs is assumed to be 10% annually from 2009 through 2014, based on the most recent two years of data, then declining to 5% growth annually in 2015 through 2020. In the IOU low case, we did not adjust growth for the economic recession because the growth rate used already mirrored the slow growth experienced in 2008.

For public utilities, we assumed growth rates were the same as for IOUs in the low case, with 10% annual growth from 2009 through 2014, falling to 5% annually in 2015 through 2020. Although we have no data to suggest this, the lower growth of IOUs might also apply to public utilities, especially as the programs get older and find it more difficult to acquire new customers.

For cooperatives, we assumed in the low case that growth in 2009 and 2010 would be consistent with 2008 observed trends, a decline of 7%. This is consistent with the assumption for cooperative programs in the high case. From 2011 through 2014, the low case assumed cooperatives would be satisfied with making the programs available to their customers and would not significantly promote them. Given this assumption of static programs and little marketing support, participation is projected to remain at 2010 levels (growth of 0%), and then decline by 5% annually from 2015 through 2020 as attrition erodes the participant base of programs.

Negative Policy Impacts Scenario

Under the negative impacts scenario, we assumed utility programs are impacted by carbon policy that eliminates green power as a carbon mitigation strategy but to a lesser extent than REC markets. Generally, utility programs are dominated by residential and small commercial customers that may not be interested in making carbon-neutral claims or participating in GHG emissions inventories, and therefore are less sensitive to the inability to do so. However, survey research suggests that some consumers are motivated to purchase green power for the environmental benefits, with one survey suggesting about 30% of consumers are concerned about long-term environmental benefits.[9] Therefore, some residential or small commercial purchasers may cease purchasing if carbon benefits are not conveyed to green power purchasers, and some marketers may cease to offer products if the environmental benefits are diminished or if the lack of carbon benefits may be considered deceptive to consumers who may think their purchases are reducing carbon emissions. In addition, a loss of a large fraction of commercial and institutional purchasers could have spillover effects on the rest of the market or result in a loss of some credibility. While significant uncertainty surrounds the actual impacts on this market segment and data with respect to the fraction of consumers motivated specifically by carbon mitigation benefits are lacking, we assumed initial growth consistent with that in the low case and a loss of one-third of existing purchases through utility programs in later years after carbon policy is adopted.

In the negative policy impacts scenario, consistent with the low case, the growth rate for existing IOU and public utility programs is assumed to be 10% annually in 2009 through 2011. For cooperatives, also consistent with the low case, we assumed a decline of 7% in 2009 and 2010. In 2011, we assumed cooperatives would see zero growth. For the 2012-2013 timeframe, we assumed all utilities would lose one-third of their 2011 demand, which is generally consistent with available survey data on customer motivations. For 2014-2020, we assumed annual decreases of 10% for all utilities.

Summary

For 2015, we forecast that existing regulated utility programs will range from 9.0 million MWh to 19.2 million MWh (Table 6). This growth represents between a 1.8- and a 3.9-fold increase from 2008 levels. The negative policy impacts scenario represents a 29% decrease from 2008, resulting in 3.5 million MWh in 2015.

5.2. New Utility Programs

Forecasting demand from new utility programs followed a slightly different approach. We determined the fraction of utility load that might offer a green power option over the forecast period, under both a high and a low case. First, we identified utilities with existing green pricing programs. These constitute nearly half of total U.S. utility retail electricity sales, as shown in Table 7.

Next, we removed all of these roughly 850 utilities with green pricing programs currently in operation, which left us with a list of utilities not currently offering a green power program in states with traditional utility regulation. Then, we estimated the portion of this load that might conceivably have a green power program made available to it over the course of our forecasting horizon, with a modest assumption of new load in the low case and a more aggressive assumption in the high case. The determination of potential new utility programs considered factors such as whether the utility has expressed interest in offering a program or is in a region where programs are already available. The new retail load to which a green option was made available was apportioned equally to each of the years of the forecast 2009-2020 in each scenario, so that some new load would be exposed to a new green power program startup in each year. Finally, we made assumptions about the percent of utility retail load served by green power in the launch year based on historical data for new programs. Subsequent year growth assumptions were informed by growth rates for existing utility programs.

High Case

In the high case, we assumed IOU load with a green power option available would increase by 25%, from 47% to 72% by 2020, while public utility and coop loads would increase by 15%, from 47% and 45% to 62% and 60%, respectively. The high case presents an optimistic view of utility interest in offering new green pricing programs. Utility interest might be increased if green pricing programs help manage regulatory risks or if utilities find the programs helpful in managing their renewable energy portfolios. The high case is also designed to reflect a higher level of utility program promotion and participating green power load, as consumer awareness of renewable energy options and interest in purchasing green power may continue to grow as it becomes more mainstream.

Low Case

In the low case, we assumed IOU load with a green power option available would increase by 10%, from 47% to 57% by 2020, while public utility and coop loads would increase by 5%, from 47% and 45% to 52% and 50%, respectively. The low case is designed to reflect modest utility interest in offering new voluntary programs in light of increasing mandates and competition for renewables to meet compliance obligations. Many utilities interested in offering green pricing programs may have already done so. Going forward, utilities may not want to offer voluntary programs that make compliance obligations more expensive and difficult to meet. Some utilities might shift to rate-basing renewables as renewable resources become more cost-competitive rather than offering green power as an optional product.

Furthermore, consumers might be less interested if the price of purchasing renewable through voluntary programs rises, or they might feel less need to support renewable energy if they believe that government is providing adequate support in the form of incentives or mandates. The low case is designed to reflect more modest participation as a because of these potential factors.

Negative Policy Impacts Scenario

Under the negative policy impacts scenario, we assumed there are no new utility programs offered. This might occur if the overall market loses credibility and if the environmental community does not support the development of new programs. The method for growing the new programs over time is the same in the low and high case.

Table 6. Historical and Projected Size of Existing Utility Green Power Programs (thousand MWh)

	2006	2007	2008	2009	2010	2011	2012	2013	2014	2015
High Case	3,400	4,300	4,900	5,600	6,600	8,200	10,200	12,800	16,000	19,200
Low Case	3,400	4,300	4,900	5,300	5,900	6,400	7,100	7,800	8,600	9,000
Negative Policy Impacts Case	3,400	4,300	4,900	5,300	5,900	6,400	5,400	4,300	3,900	3,500

Table 7. Percent of Utility Retail Sales with Green Power Option Currently Available

IOUs	Public Utilities	Cooperatives
47%	47%	45%

Utility green power programs see a substantial "bump" in sales in the first and second years of the program because programs are often heavily promoted when they are initially launched and the early adopters enroll in the program. We assumed participation is 0.15% of the load newly exposed to a green power option in the first year based on historical green pricing program data from NREL (Table 8).[10]

In the second year, we assumed these new programs grow by 140%, again based on average data for growth of green pricing programs in their second year. Because of limited data on second year growth for programs that began in 2004 (3 observations), we used an average of programs that started in 2002 and 2003 (Table 9). For subsequent years, we used an average growth rate for existing utility programs, differentiated by utility type. We used the same rates for the high case, low case, and negative policy scenario. Specifically, we assumed IOU and public utility programs would grow at an average annual rate of 15%, and that cooperative programs would remain constant (0% growth) after the initial first and second years of growth. The growth rates used in the third and subsequent years are between the high- and low-growth rates used for forecasting the size of existing green pricing programs (see Section 5.1); the rates were selected to reflect new program growth without the bump in the first couple of years of the program.

Table 8. Average Green Power Load during First Year of Program

Program-Year Start	Average Percent of Green Power Load for Programs Starting in Year
2003	0.11%
2004	0.09%
2005	0.12%
2006	0.02%
2007	0.15%
2008	0.36%
Average	0.14%

Table 9. Utility Green Pricing Programs Second Year Growth

Program-Year Start	2nd Year Growth	3rd Year Growth	4th Year Growth	5th Year Growth
2002	129%	96%	36%	27%
2003	149%	11%	11%	10%
Average	139%	53%	23%	19%

Table 10. New Utility Programs (thousand MWh)

	2009	2010	2011	2012	2013	2014	2015
High Case	70	250	460	690	960	1,260	1,600
Low Case	30	100	180	270	370	480	620
Negative Policy Impacts Case	-	-	-	-	-	-	-

Summary

New regulated utility programs are expected to reach between 620,000 MWh and 1.6 million MWh by 2015 (Table 10), under the low and high cases respectively. The negative policy impacts scenario results in no new utility programs.

6. Restructured Markets

In states that have restructured electricity markets, green power options take a different form than those offered by traditionally regulated utilities. Therefore, we treated states with retail competition in the electric sector differently in this analysis. The voluntary demand for renewable energy in restructured markets is highly dependent on 1) changes in wholesale market prices, 2) requirements by states that standard-offer or default suppliers offer a green power option, and 3) the facilitation by states or utilities of green options available through customer electricity bills.

Fourteen states with restructured electricity markets allow retail customers to choose their electricity provider.[11] We do not anticipate any additional states will support customer choice at the retail level during the forecast horizon. In the 14 states with restructured markets, green power options have generally been presented to consumers either (1) as a bundled electricity product in which customers switch from standard offer providers to green power providers or (2) as an unbundled REC add-on to default electricity service, which we refer to here as "competitive programs." Offerings in the latter category are referred to as programs because they tend to be green power options sanctioned by state public utility commissions in conjunction with the provision of default supply.

Initially, competitive electricity providers offered a green power product bundled with electricity in many of these restructured electricity markets, but many of these bundled green power products were dropped because of difficulties competing in the market, and because in residential markets at least, customer switching to new providers was never very strong. Nevertheless, several of these competitive electricity providers continue to market green power products to retail customers. We treated them separately from "competitive programs" because of differences in their promotion, in their uptake rates, and in available historical participation data.

6.1. Competitive Marketers

In states such as Illinois, Maine, Maryland, New York, Pennsylvania, and Texas, green electricity products are marketed by electricity service providers. NREL has historical data from 2004 on competitive market sales that are based on information provided by marketers, but NREL does not have data for individual states. These markets have experienced overall growth over the past several years, but the growth has been mixed with declines as some marketers have exited markets (e.g., Connecticut, Ohio, and Pennsylvania) when market rules or conditions have made it too difficult to offer green products. Nevertheless, competitive market sales have experienced compound annual average growth rates of 4% for 2004-2008 and 13% for 2005-2008 (Table 11). These rates do not include growth in competitive programs, which are addressed in Section 6.2.

High Case and Low Case

For 2009-2010, we forecast that competitively marketed green power would grow by 5% per year in the high case and 0% in the low case due to the economic downturn. For 2011-2014, based on historical growth rates, we assumed demand for competitively marketed green power would continue to grow, in the high case by 15% per year, and in the low case by 5% per year. For 2015-2020, we assumed, as we have with the other market segments, that the market would mature and that attracting customers would be more difficult. Therefore, growth is estimated to slow in the high case to 10% annually and in the low case, 0%.

The high case reflects a more optimistic growth scenario, assuming that marketers can continue to gain traction and compete against wholesale prices in existing restructured markets. It also assumed relatively favorable market conditions and market rules persist. The low case assumed slower growth, reflecting more significant challenges to remaining competitive in these markets in the future.

Negative Policy Impacts Scenario

In the negative policy impacts scenario, we assumed declines similar to those of the REC market because this market segment is also dominated by generally large corporate or institutional purchasers and the same issues apply (see earlier discussion in Section 0). We assumed no growth in 2009 through 2011 and then a loss of two-thirds of 2011 demand by 2013. From 2014 onward, we assumed annual decreases of 10%.

Summary

Competitive marketers are expected to represent between 3.6 million MWh and 6.4 million MWh in 2015 (Table 12). This represents growth from 2008 of 1.2-fold in the low case to 2.1-fold in the high case. Compared to growth in other sectors, this growth is relatively modest. The negative policy impacts scenario represents a 73% decrease from 2008, resulting in 0.8 million MWh in 2015.

Table 11. Historical Growth in Competitive Marketers Sector

Year	Competitive Marketers (thousand MWh)	Change from Previous Year
2004	2,513	-
2005	1,861	-26%
2006	1,715	-8%
2007	2,532	48%
2008	2,996	18%
CAGR 2004-2008		4%
CAGR 2005-2008		13%

6.2. Existing Competitive Programs

Competitive programs (also sometimes called “default supplier” or “check-off programs” for the ease in which customers can enroll) are currently available in several states, including Connecticut, Massachusetts, New York, New Jersey, and Rhode Island. These are programs in which REC marketers offer specific products that can be made available either centrally, such as through a state-sponsored Web site or through the Web site of a default service provider (i.e., usually through the distribution utility). We categorized these as competitive programs rather than as REC sales because the RECs are essentially bundled with the customer's electricity and the customer pays only one bill. It is up to individual marketers to promote their own products, but billing and collections are handled by the utility. Historically, competitive programs grew an average of 48% annually between 2005 and 2008, though the annual growth rate from 2007 to 2008 was only 25% (Table 13).

High Case

To forecast continued demand for these programs, we assumed demand would grow in the high case by 25% from 2008-2009 and 2009-2010, and then 30% per year for 2011-2014 period. We assumed that this growth rate cannot be sustained through 2020, and therefore reduce growth rates to 25% in 2015 and 2016, 20% in 2017 and 2018, and 15% in 2019 and 2020.

The slower growth assumption for 2008-2010 is intended to reflect the effects of the economic downturn, although the impacts of the economy are assumed to be more modest in the high case than in the low case. The rate is based on the growth rate for existing programs of this type during 2008. The growth assumptions for the later years are more conservative than the historic rates, but are designed to reflect the challenges in continuing to attract consumer segments after the most motivated customers are already enrolled. The high case generally reflects an optimistic scenario of growth for these programs going forward, assuming that they are actively promoted and that consumers are favorably disposed to participate. These programs have grown more rapidly than green pricing programs in regulated markets largely because these competitive providers do not make money unless they increase their sales.

Low Case

In the low case, we assumed sales grow at 10% during 2009 and 2010, because of the recession, and then grow at an annual rate of 15% for 2011-2014. Similar to the high case, we assumed growth rates will begin to decline in 2015. Growth rates were reduced to 10% in 2015 and 2016, 5% in 2017 and 2018, and 0% in 2019 and 2020.

The more modest growth rate reflects a scenario in which marketers do not place as much attention and resources in growing these programs as they have in the past. Also, it assumes that states do not continue to provide incentives or support for developing these markets, as has been the case in the past and has likely fueled some of the higher growth rates. The more modest growth also reflects decreased interest on the part of consumers in participating in these kinds of programs as a result of policy considerations noted earlier and perhaps higher renewable energy prices.

Table 12. Competitive Marketers (thousand MWh)

	2004	2005	2006	2007	2008	2009	2010	2011	2012	2013	2014	2015
High Case	2,500	1,900	1,700	2,500	3,000	3,100	3,300	3,800	4,400	5,000	5,800	6,400
Low Case	2,500	1,900	1,700	2,500	3,000	3,000	3,000	3,100	3,300	3,500	3,600	3,600
Negative Policy Impacts Case	2,500	1,900	1,700	2,500	3,000	3,000	3,000	3,000	2,000	1,000	900	800

Table 13. Growth Rates for Existing Competitive Program Sector

Year	Competitive Markets (MWh)	Growth from previous year
2005	290,780	139%
2006	491,243	69%
2007	758,296	54%
2008	946,586	25%
CAGR 2005-2008		48%

Negative Policy Impacts Scenario

In the negative policy impacts scenario, we assumed reductions in default supplier programs in competitive markets similar to those for regulated utility programs, because they are also dominated by residential and small commercial customers. Declines in these programs could occur if carbon policy is adopted that results in significant losses of large commercial and institutional REC purchasers, which result in spillover effects in other market segments and a potential loss of credibility in the market. Also, consumers interested in carbon benefits may cease purchasing as discussed earlier (see Section 0). We assumed growth in 2009 and 2010 is 10% annually, and growth in 2011 is 15%, consistent with the low case. In 2012-2013, one-third of 2011 demand is lost. From 2014-2020, 10% of demand is lost annually.

Table 14. Projected Size of Existing Competitive Programs (thousand MWh)

	2008	2009	2010	2011	2012	2013	2014	2015
High Case	900	1,200	1,500	1,900	2,500	3,200	4,200	5,300
Low Case	900	1,000	1,100	1,300	1,500	1,700	2,000	2,200
Negative Policy Impacts Case	900	1,000	1,100	1,300	1,100	900	800	700

Summary

Existing competitive programs (default supplier programs) are forecasted to be between 2.2 million MWh in the low case and 5.3 million MWh in the high case (Table 14). This represents between a 2.3-fold and a 5.6-fold increase from 2008. The negative policy impacts scenario represents a 25% decrease from 2008, resulting in 0.7 million MWh in 2015.

6.3. New Competitive Programs

Other states with restructured electricity markets could act over the next several years to encourage voluntary green power options. Most existing competitive programs emerged because of action by utility regulators or legislative action, and they were often driven by support from environmental and consumer protection organizations. For example, the Maine legislature passed an energy bill in 2009 that requires the Maine Public Utilities

Commission to provide a green power option to all residential and small commercial electricity customers, in addition to the standard offer or default service. Maine's green power option or options will be selected through a competitive process and are expected to be available to consumers in 2011.

While we have not attempted to predict what form these program options may take, we made assumptions about the load that will be exposed to a green option and then forecast market size using a similar approach as was used for the new electric utility programs (see Section 5).

High Case

For the high case, we assumed six restructured states (New Hampshire, Pennsylvania, Maryland, Delaware, Ohio, and Illinois) that currently do not have a competitive program, in addition to Maine, would require that such a program be offered in the future. Given that Maine's offering is still being developed and that no other states have taken action to require a green power offering to date, we assumed no new offerings would come online in 2009 or 2010. Thus, we assumed, for the high case, that all IOU load in these seven states would be exposed to a green power option over the course of the forecast, 2011-2020.

We assumed these programs would experience an uptake of 0.15% of retail load, in the first year, similar that of new utility programs, and an increase of 150% in participation from year one to year two as the most strongly supportive customers sign up. Beyond the second year through 2015, we further refined the high case to assume that demand would grow at an annual rate of 50%, reflecting a scenario in which there is effective promotion by the providers and state support for customer awareness. From 2015 through 2020, we assumed growth of 20% annually, which reflects the challenges of attracting new customers as programs mature, and generally consistent with growth assumed for the existing competitive programs over that period.

Low Case

For the low case, we assumed Maine and two other states would require green options over the forecast period. Therefore, we assumed the IOU load in Maine and two-sixths of the IOU load in the other six states would be exposed to a green power option. This way we did not have to make any further assumptions about exactly which states would be most likely to create a new competitive program. As with the high case, we assumed no new programs would be added in 2009 or 2010. In the 2011-2020 timeframe, we assumed programs would experience an uptake of 0.15% of retail load, in the first year, similar that of new utility programs, and an increase of 150% in participation from year one to year two as the most strongly supportive customers sign up. From years three to 2015, we assumed annual growth would be 15% in the low case, based on a scenario of modest promotion and little educational support due to a lack of funding. From 2015-2020, we assumed growth slows to 10% annually, generally consistent with growth assumed for the existing competitive programs over that period.

Negative Policy Impacts Scenario

In the negative policy impacts scenario, we assumed only Maine would require a green option over the forecast period, because legislation requiring it has already been adopted. Consistent with our assumptions about utility programs, we assumed no other new default supplier programs would be offered over the forecast period because of a potential loss of credibility and overall decline in the market. We assumed Maine's offering would begin in 2011. In the initial year of the program, we assumed only 0.10% of retail load would participate in the green power offering, and that there would be an increase of 100% in participation from year one to year two. From years three to 2015, we assumed more modest growth of 10% annually, and in 2015-2020, we assumed negative annual growth of 10%.

Summary

In restructured markets, new competitive programs are forecasted to represent 360,000 MWh in the low case and 1.35 million MWh in the high case (Table 15). The negative policy impacts scenario forecasts 10 thousand MWh in 2015.

7. On-site Renewable Energy

Demand for on-site renewable energy is expected to grow significantly over the next 5 to 10 years because of declining solar prices and state incentives. To forecast demand for on-site renewables, we developed separate projections for solar photovoltaics (PV) and other on-site resources. We relied on existing market projections available in the literature for the on-site solar PV forecast. The other on-site renewables were projected based on historical data from the EPA Green Power Partnership.[12]

For on-site renewable energy, we did not develop alternate assumptions for the negative policy impacts scenario because growth of on-site generation is not likely to be negatively impacted by future carbon policy, as it is driven primarily through reductions in cost as well as consumer interest in owning renewable generation. Organizations installing on-site systems would reduce their electricity purchases, and would likely be able to continue to reflect the purchased electricity emissions reductions in GHG accounting.[13] Also, carbon policy would likely help improve the cost-effectiveness of on-site systems. Therefore, the assumptions for on-site in this analysis are the same in the low case and the negative policy impacts scenario.

Table 15. New Competitive Programs (thousand MWh)

	2009	2010	2011	2012	2013	2014	2015
High Case	0	0	80	280	570	920	1,350
Low Case	0	0	30	100	170	260	360
Negative Policy Impacts Case	0	0	1	3	6	8	10

7.1. On-site Solar Projections

High and Low Case

For solar, we used projections that Navigant Consulting (Navigant) recently completed for NREL (Paidipati et al. 2008) to develop a low case and a high case. The Navigant study included four scenarios: worst case, base case, focused policies, and base case (Table 16).

We selected the focused policies case for both our low and high cases because it included an assumption that the federal tax credit would be fully extended through 2016 for both commercial and residential customers, and because it lifted the $2,000 cap on residential tax credit. Because this policy has already been enacted, the worst-case and base-case scenarios were not used in our analysis.

The focused policies scenario also assumed that net metering is nationally available, that net metering caps are lifted (these improvements had a large impact, a 58% increase over the base case), and that price for carbon is $12.00 in 2012 and rises to $13.89 in 2015. The best-case scenario was not used because it was more aggressive than other forecasts that we examined and because of the potential challenges in achieving the assumptions. Table 16 compares the assumptions in each scenario.

Table 16. PV Scenario Assumptions in Navigant Study

Scenario	Worst Case	Base Case	Focused Policies	Best Case
Interconnection Policy Scenario	current rules	current rules	current rules	improved
Net Metering Availability Scenario	current availability	current availability	nationwide availability	nationwide availability
Net Metering Cap Scenario	current caps	current caps	caps lifted	caps lifted
Cap and Trade Scenario	Low Carbon Economy Act of 2007	none	Low Carbon Economy Act of 2007	Low Carbon Economy Act of 2007
Electricity Price Escalation	Energy Information Administration (EIA) projections	accelerated	accelerated	accelerated
Federal Tax Credit	baseline	extended	fully extended	fully extended
Time-of-Use Rates	current availability	current availability	current availability	nationwide availability
RPS Solar Set-Aside Enforcement	no	yes	yes	yes

Source: Paidipati et al. (2008).

Navigant produced two forecasts of solar prices: a business as usual case (BAU) and a Solar America Initiative case, which assumed more aggressive cost reductions (Table 17). These cost reduction scenarios were used to create our low and high cases. We selected the focused policy scenario with aggressive price reductions (DOE Solar America Initiative pricing) for our high case and the focused policy scenario with business as usual price reductions for our low case.

Because the Navigant analysis projects solar generation for all purposes, we subtracted from these projections the amount of solar energy required by state solar carve-outs—as calculated by Lawrence Berkeley National Laboratory—so as not to double count generation for both the voluntary and the compliance markets. We did not account for any on-site systems in which RECs may be sold for RPS compliance other than through solar set-asides. We did not attempt to address this issue because of the difficulties in estimating the potential magnitude. However, it is only likely to occur in areas (such as California) in which PV is heavily subsidized—thus, making solar competitive with other types of renewable generation for RPS compliance—or if future cost reductions make it competitive.

The Navigant solar PV scenarios only project the growth of solar PV only to 2015. We extended the projections to 2020 by applying a growth rate calculated from two Solar Deployment System model scenarios (Denholm et al. 2009). We selected two model runs from the Denholm et al. (2009) study that used DOE Solar Energy Technology Program PV module cost projections and the market share estimates from the Navigant study (Paidipati et al. 2008). For our low case in the "out" years, we used a reference case with no price on carbon, and for our high case, we used an attractive case, assuming 1% annual electricity rate escalations and carbon prices between $20/ton and $30/ton through 2020. Solar electric demand for RPS compliance was subtracted from the two projections, and annual growth rates were calculated. The annual growth rates from the reference case were applied to the low-case projection and the annual growth rates from the attractive case were applied to the high-case projection. Based on these assumptions, we projected voluntary market demand in 2015 for on-site solar would be 11.5 million MWh in the low case and 21.4 million MWh in the high case (Table 18). This growth represents a 4.7-fold increase in the low case and an 8.4-fold increase in the high case from 2009 levels.

Table 17. Navigant Study Price Reduction Assumptions

Price Projections		Retrofit Installed System Price ($2007)			New Construction Installed System Price ($2007)		
		2007	**2010**	**2015**	**2007**	**2010**	**2015**
Business as Usual	Residential	$7.40	$6.20	$4.80	$7.40	$5.90	$4.50
	Commercial	$6.41	$5.80	$4.50	$6.70	$5.50	$4.20
Solar America Initiative	Residential	$7.40	$5.11	$3.10	$7.10	$3.86	$2.44
	Commercial	$6.41	$3.75	$2.49	$6.23	$3.60	$2.32

Source: Paidipati et al. (2008).

Table 18. On-site Solar (thousand MWh)

	2009	2010	2011	2012	2013	2014	2015
High Case	2,600	3,400	5,000	7,200	9,700	15,400	21,400
Low Case	2,400	2,800	3,900	5,400	6,400	9,400	11,500

7.2. Other On-site Renewables

For non-solar on-site generation, we used data from the EPA's Green Power Partnership showing non-residential demand for on-site resources broken down by resource type. Most of this demand is being met by biogas, biomass, and to a lesser extent solar, wind and hydroelectric projects, as shown in Table 19.

Table 19. Generation by Resource (MWh), 2009

	Total Generation (MWh)	Resource Breakdown
Biogas	411,301	52.52%
Biomass	203,923	26.04%
Geothermal	0	0.00%
Hydroelectric	18,223	2.33%
Solar	95,470	12.19%
Unknown	0	0.00%
Wind	54,165	6.92%
Total	783,083	--

Source: EPA (2009).

High and Low Case

Using historical data from EPA for voluntary support for non-solar, on-site generation (shown in Table 19), we assumed 35% annual growth in the high case and 15% annual growth in the low case to reflect a more modest growth scenario through 2015. To account for potential limitations in availability of biomass available for on-site generation and to account for the fact that growth rates often moderate over time, we lowered the non-solar, on-site growth rates to 25% (high case) and 10% (low case) for 2016-2020.

In 2015, other on-site renewables are projected to be between 1.8 million MWh and 4.8 million MWh (Table 20). From 2009, this represents 2.3-fold growth in the low case and 6-fold growth in the high case.

Summary of Results

Given the assumptions used in each market sector, our forecast projects the total voluntary demand for renewable energy in 2015 to range from 63 million MWh annually in the low-case scenario to 157 million MWh annually in the high-case scenario (Figure 2). The range represents an approximately 2.5-fold difference. The negative policy impacts scenario reflects a market size of 24 million MWh.

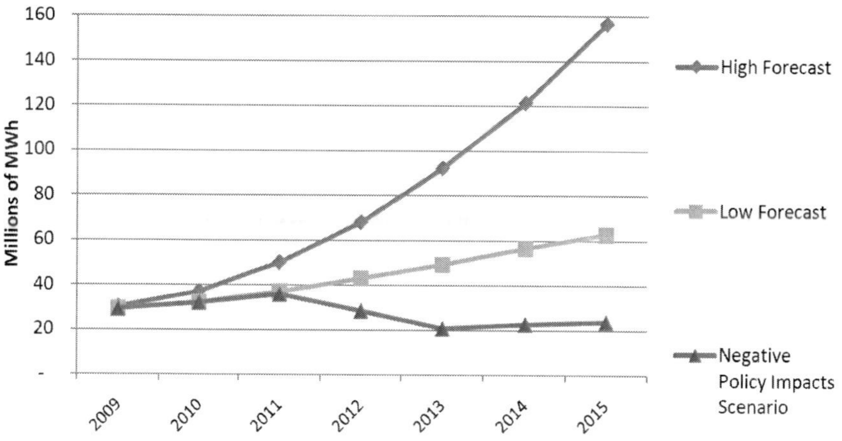

Figure 2. Total voluntary demand for renewable energy (millions of MWh): High, low, and negative policy impacts scenario forecasts through 2015.

Table 20. Other On-Site Renewables (thousand MWh)

	2009	2010	2011	2012	2013	2014	2015
High Case	800	1,100	1,400	1,900	2,600	3,500	4,800
Low Case	800	900	1,000	1,200	1,400	1,600	1,800

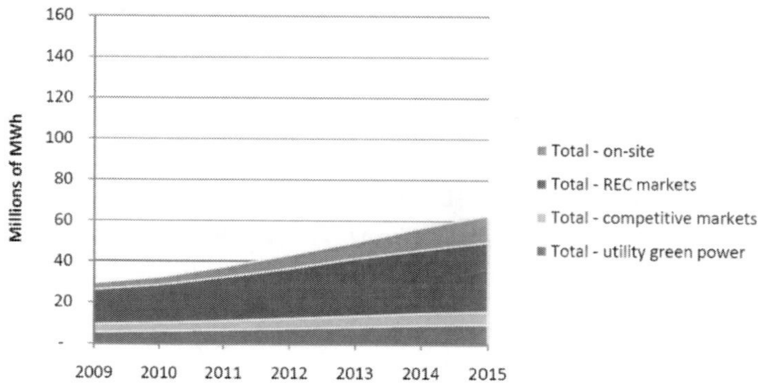

Figure 3. Low forecast of total voluntary demand for renewable energy (millions of MWh) by sector, through 2015.

The forecast uses historical data, including market sector sizes and growth rates, to inform our assumptions. While growth rates have been adjusted as discussed in the previous sections, they remain the starting point for our analysis. The difference between the low and high cases reflects market and policy uncertainty and the difficulty predicting how consumers will react to various factors affecting the voluntary market. The negative policy impacts scenario is designed to reflect impacts on the market that might occur if future carbon policies significantly reduce the motivation among large purchasers in particular. This scenario reflects the potential loss of a significant volume of existing purchasers.

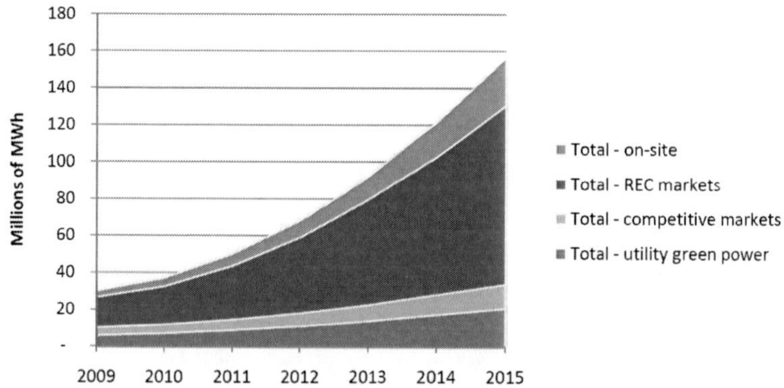

Figure 4. High forecast of total voluntary demand for renewable energy (millions of MWh) by sector, through 2015.

In 2008, total voluntary demand represented 0.65% of total retail electricity sales in the United States. To provide some perspective on future estimates, we compare them to total projected retail electricity sales in the United States. The high case represents 3.8% of total retail electricity sales projected by the DOE Energy Information Administration (EIA) in its Annual Energy Outlook 2009 (EIA 2009), while the low and negative-impacts cases represent 1.5% and 0.6% of total sales, respectively. In addition, a recent analysis that estimated ranges of renewable energy supplies in 2015 (Bird et al. 2009b), shows that under a high supply scenario, these demand levels could be achieved.[14]

In both the high and low forecasts, REC markets continue to drive much of the growth. In 2008, REC markets represented nearly two-thirds of the total voluntary market. In 2015, REC markets are forecasted to represent between 54% and 62% of the total market. However, in the negative policy impacts scenario, the contribution of RECs in 2015 is significantly less, at 23% of the total market.

On-site solar is projected to increase its share of the total voluntary market. In 2009, onsite solar is estimated to represent 8-9% of the total voluntary market, whereas in 2015, on-site solar is forecasted to represent between 14% and 18% of the total market. In the negative policy impacts scenario, on-site solar represents 49% of the total market in 2015.

Utility green power programs are projected to decline as a share of the total voluntary market. In 2008, utility programs represented approximately 20% of the total voluntary market. By 2015, existing and new utility programs together are forecasted to represent between 13% and 15% of the total market. In the negative policy impacts scenario, existing and new utility programs represent 15% of the total market, although no new utility programs are forecasted.

The competitive market, while continuing to increase in size, represents less of the total voluntary market in 2015 than in 2008. In 2008, the competitive market represented 16% of the total voluntary market, but by 2015 it is forecasted to represent only between 8% and 10% of the total voluntary market. In the negative policy impacts scenario, the competitive market represents 6% of the total market in 2015.

Voluntary demand through 2015 was the focus of this forecast, but projections were extended through 2020. The 2020 projections show 327 million MWh annually in the high case and 94 million MWh annually in the low case. This range is greater than the 2015 range, representing an approximately 3.5-fold difference between high and low cases. The negative policy impacts scenario projects 33 million MWh in 2020.

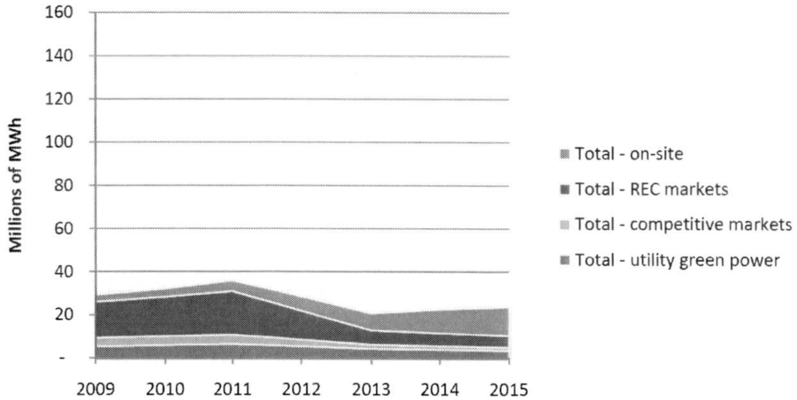

Figure 5. Negative policy impacts scenario forecast of total voluntary demand for renewable energy (millions of MWh) by sector, through 2015.

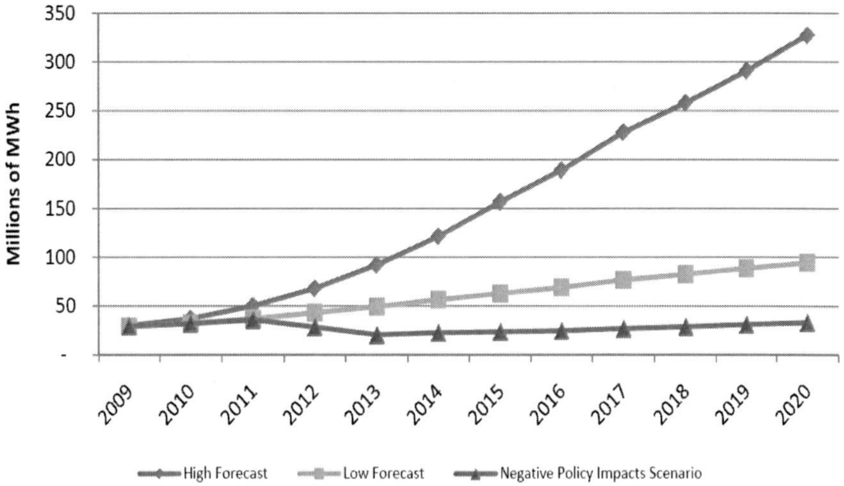

Figure 6. Total voluntary demand for renewable energy (millions of MWh): High, low, and negative policy impacts scenario forecasts through 2020.

Uncertainties

A number of key uncertainties affect the results of this forecast.

- The growth assumptions are informed largely by historical growth rates in each submarket, but it is unclear where the market lies along the S-curve of market diffusion and how consumer interest will change.
- There is significant policy uncertainty over the time period analyzed. The design details and timing of any potential federal and state RPS policies as well as regional or federal cap and trade programs will have implications for the market that are difficult to predict without specific knowledge of the policies.
- Uncertainty surrounding the price of renewable generation over the period analyzed and its relative competitiveness with conventional generation has implications for consumer willingness to pay for renewables. Voluntary market demand can be price sensitive.
- There is significant uncertainty in the level of interest that utilities will show in offering and promoting green power products going forward. The level of promotion is an important factor in determining participation.

Appendix. Summary Results

Table A-1. Projected Size of Voluntary Market, High Case (thousand MWh)

	2009	2010	2011	2012	2013	2014	2015
Existing utility programs	5,600	6,600	8,200	10,200	12,800	16,000	19,200
New utility programs	100	300	500	700	1,000	1,300	1,600
Existing competitive programs	1,200	1,500	1,900	2,500	3,200	4,200	5,300
New competitive programs	-	-	80	280	570	920	1,350
Competitive marketers	3,100	3,300	3,800	4,400	5,000	5,800	6,400
REC markets	16,700	20,800	29,100	40,800	57,100	74,300	96,500
On-site non-solar	800	1,100	1,400	1,900	2,600	3,500	4,800
On-site solar	2,600	3,400	5,000	7,200	9,700	15,400	21,400
TOTAL	**30,100**	**37,000**	**49,980**	**67,980**	**91,970**	**121,420**	**156,550**

Table A-2. Projected Size of Voluntary Market, Low Case (thousand MWh)

	2009	2010	2011	2012	2013	2014	2015
Existing utility programs	5,300	5,900	6,400	7,100	7,800	8,600	9,000
New utility programs	30	100	180	270	370	480	620
Existing competitive programs	1,000	1,100	1,300	1,500	1,700	2,000	2,200
New competitive programs	-	-	30	100	170	260	360
Competitive marketers	3,000	3,000	3,100	3,300	3,500	3,600	3,600
REC markets	16,700	18,300	21,100	24,200	27,900	30,600	33,700
On-site non-solar	800	900	1,000	1,200	1,400	1,600	1,800
On-site solar	2,400	2,800	3,900	5,400	6,400	9,400	11,500
TOTAL	**29,230**	**32,100**	**37,010**	**43,070**	**49,240**	**56,540**	**62,780**

Table A-3. Projected Size of Voluntary Market, Negative Policy Impacts Scenario (thousand MWh)

	2009	2010	2011	2012	2013	2014	2015
Existing utility programs	5,300	5,900	6,400	5,400	4,300	3,900	3,500
New utility programs	-	-	-	-	-	-	-
Existing competitive programs	1,000	1,100	1,300	1,100	900	800	700
New competitive programs	-	-	1	3	6	8	10
Competitive marketers	3,000	3,000	3,000	2,000	1,000	900	800
REC markets	16,700	18,300	20,200	13,400	6,700	6,000	5,400
On-site non-solar	800	900	1,000	1,200	1,400	1,600	1,800
On-site solar	2,400	2,800	3,900	5,400	6,400	9,400	11,500
TOTAL	**29,200**	**32,000**	**35,801**	**28,003**	**19,106**	**21,808**	**23,710**

REFERENCES

Bird, L., Holt, E. & Carroll, G. (2007). "*Implications of Carbon Regulation for Green Power Markets.*" NREL/TP-640-41076. Golden, CO: National Renewable Energy Laboratory. http://apps3.eere.energy. gov/greenpower/ resources/pdfs/41076.pdf

Bird, L., Kreycik, C. & Friedman, B. (2009a). "*Green Power Marketing in the United States: A Status Report (2008 Data).*" NREL/TP-6A2-46581. Golden, CO: National Renewable Energy Laboratory. http://apps3. eere.energy.gov/greenpower/resources/ featured.shtml

Bird, L., Hurlbut, D., Donohoo, P., Cory, K. & Kreycik, C. (2009b). "*Examination of the Regional Supply and Demand Balance for Renewable Electricity in the United States through 2015.*" NREL/TP-6A2-45041. Golden, CO: National Renewable Energy Laboratory. http:// www.nrel.gov/ docs/fy09osti/45041.pdf

Bird, L. & Lokey, E. (2007). "*Interaction of Compliance and Voluntary Renewable Energy Markets.*" NREL/TP-670-42096. Golden, CO: National Renewable Energy Laboratory. http://apps3.eere.energy.gov/ greenpower/pdfs/42096.pdf

Bird, L. & Kaiser, M. (2007). "*Trends in Utility Green Pricing Programs (2006).*" NREL/TP-670-42287. Golden, CO: National Renewable Energy Laboratory. http://apps3.eere.energy.gov/ greenpower/pdfs/42287.pdf

Busch, C. (2010). Center for Resource Solutions Comments to Western Climate Initiative on "*Voluntary Renewable Energy Market: Issues and Draft Recommendations.*" 19 February. http://www.westernclimate initiative.org/components/com_publiccomments/comments_documents/66 6/Center_for_Resource_Solutions_on_VRE.pdf.

Denholm, P., Drury, E. & Margolis, R. (2009). "*The Solar Deployment System (SolarDS) Model: Documentation and Sample Results.*" NREL/TP-6A2-45832. Golden, Colorado: National Renewable Energy Laboratory. http://www.nrel.gov/docs/fy10osti/45832.pdf

Energy Information Administration (EIA). (2009). "*Annual Energy Outlook 2009 with Projections to 2030.*" DOE/EIA-0383(2009). http:// www.eia.doe.gov/oiaf/aeo/pdf/0383 (2009).pdf

Harmon, R. & Hirschhorn, M. (2006). "Clearing the Air: The Impact of Carbon Cap and Trade Regulations on Renewable Energy." *SOLAR 2006 Conference Proceedings, July 2006.*

Holt, E. & Wiser, R. (1999). *Understanding Consumer Demand for Green Power.* National Wind Coordinating Committee.

Holt, E., Wiser, R., Fowlie, M., Mayer. R. & Innis, S. (2001). Understanding Non-Residential Demand for Green Power. *National Wind Coordinating Committee.*

Natural Marketing Institute (NMI). (2009). *Consumer Trends & Profiles.* Presentation at the Renewable Energy Markets 2009 Conference. September 14.

Opinion Research Corporation. (2006). *Renewable Energy Survey*, Study #715268, June 29, 2006. Prepared for National Renewable Energy Laboratory, Golden, Colorado.

Paidipati, J., Frantzis, L., Sawyer, H. & Kurrasch, A. (2008). "*Rooftop Photovoltaics Market Penetration Scenarios.*" NREL/SR-581-42306. http://www.nrel.gov/docs/ fy08osti/42306.pdf

Wiser, R., Bolinger, M., Holt, E. & Swezey, B. (2001). "*Forecasting the Growth of Green Power Markets in the United States.*" NREL/TP-620-30101, Golden, CO: National Renewable Energy Laboratory. http:// apps3.eere.energy.gov/greenpower/resources/pdfs/ 30101.pdf

U.S. Environmental Protection Agency (EPA). (2006, November). EPA's Green Power Partnership: An Environmental Choice for Your Organization. EPA-430-F-06-105.

U.S. Environmental Protection Agency (EPA). (2009). Nonresidential purchase data from EPA Green Power Partnership.

U.S. Environmental Protection Agency (EPA). (2010). Nonresidential purchase data from EPA Green Power Partnership.

End Notes

[1] "Green power is electricity generated from environmentally preferably renewable resources, such as solar, wind, geothermal, low-impact biomass and low-impact hydro resources" (EPA 2006).

[2] This scenario assumes that consumers continue to purchase green power beyond that required by RPS targets (e.g., they may want to purchase renewable energy equivalent to 100% of their usage). An earlier NREL study found "no evidence to suggest that the adoption of RPS policies has adversely affected voluntary purchases of renewable energy to date" (Bird and Lokey 2007).

[3] This might be the case if a set-aside for voluntary renewable energy markets is adopted in future cap and trade programs, such as in the case of the Regional Greenhouse Gas Initiative (RGGI), the first operating cap and trade program in the United States, launched in 2009 by Northeast and Mid-Atlantic states. RGGI includes a voluntary renewable energy market set-aside that preserves the carbon benefits of voluntary green power purchases by retiring an equivalent number of carbon allowances. A similar provision has been included in the California preliminary draft cap and trade rules. Whether other emerging regional cap and trade programs or potential federal programs will include such provisions is unclear.

[4] There is significant uncertainty regarding this issue, as carbon policies are currently under development and may or may not be designed to preserve voluntary renewable energy purchasers' carbon claims. Because renewable energy sources create avoided emissions benefits by displacing fossil fuel generation, renewable energy generation may not be able to affect emissions levels once a carbon emissions cap is in place. Under some carbon policy designs, renewable energy sources would make it easier for emitters to achieve compliance with an emissions cap and would enable them to sell allowances to other emitters when renewable energy displaces fossil generation. Other policy designs are possible as well.

[5] While NREL and EPA data tend to be fairly consistent, NREL REC sales numbers are generally 1-3 million kilowatt-hours (kWh) larger than the EPA data; this is because the EPA data capture only organizations that have joined its Partnership, whereas the NREL data may capture nonparticipants. For this reason, we added the incremental new 2009 REC purchases by organizations in the U.S. EPA Green Power Partnership to the 2008 NREL REC market estimate to determine total REC market demand in 2009, which is the basis for forward projections (see Table 4).

[6] Most RECs are purchased by non-residential consumers. A 2000 survey of 464 nonresidential customer purchasing green power found that their strongest motivation was fulfilling their organization's "strong and pervasive commitment to public health and the environment" (Holt et al. 2001). This research does not highlight global warming concerns but does indicate that the respondents were internally driven by their own environmental goals. Some surveys have found that residential purchasers are interested in renewable energy because of environmental benefits, but none has addressed carbon benefits specifically. In a 2009 survey, 46% of respondents indicated their top reason for supporting renewable energy was "protecting the environment" (NMI 2009). In a 2006 poll, 32% of respondents purchased or wanted to purchase renewable energy to improving the environment and 30% were

concerned with leaving their children and grandchildren a cleaner environment (Opinion Research Corporation 2006). However, the nonresidential customers—not these residential customers—are primarily driving the demand for RECs.

[7] Federal agencies have purchased RECs as well as other forms of green power to fulfill federal requirements to purchase renewable energy. The Energy Policy Act of 2005, requires federal agencies to procure renewable energy for 7.5% of electricity usage by 2013 and thereafter. Total federal facility electricity usage in 2008 was approximately 56 million MWh.

[8] It should be noted that the cooperative green power sales data included one large generation and transmission cooperative that controls substantial renewable generation and consequently has sales to a number of large wholesale customers. Because this one program accounted for several times the sales of the remaining cooperatives and skewed the results for the category as a whole, we adjusted the cooperative data shown in Table 5 by removing this program. We have assumed this cooperative program would grow at a rate consistent with the public utility average growth rate.

[9] Some surveys have found that residential purchasers are interested in renewable energy because of environmental benefits, but none has specifically addressed carbon benefits. In a 2009 survey, 46% of respondents indicated their top reason for supporting renewable energy was "protecting the environment" (NMI 2009). In a 2006 poll, 32% of respondents purchased or wanted to purchase renewable energy to improve today's environment, and 30% were concerned with leaving their children and grandchildren a cleaner environment (Opinion Research Corporation 2006).

[10] NREL has data on the green power sales for at least 35 programs in the year of their launch.

[11] According to the Energy Information Administration, the following states have restructured electricity markets: Connecticut, Delaware, Illinois, Maine, Maryland, Massachusetts, Michigan, New Hampshire, New Jersey, New York, Ohio, Pennsylvania, Rhode Island, and Texas. Other states once supported retail choice but have since suspended that policy.

[12] NREL estimates of the current green power market do not include on-site renewables; however, the U.S. EPA Green Power Partnership has data on nonresidential purchases of on-site renewables in recent years.

[13] On-site system owners could likely continue to report reductions in Scope 2 emissions (those associated with electricity purchases) in GHG inventories because they purchase less electricity. However, organizations that installed on-site systems would also not be able to make carbon emission reduction claims under some cap and trade program designs.

[14] A recent analysis of potential renewable energy supplies (Bird et al. 2009b) found that in excess of 150 million MWh of renewable energy generation could be available in excess of what is needed to meet RPS requirements in 2015 under a high wind scenario, which would be adequate to meet the high case of voluntary renewable energy demand estimated in this analysis.

In: Voluntary Markets in Renewable Energy ISBN: 978-1-62417-561-9
Editor: Jamir S. Fuller

Chapter 4

EPA'S GREEN POWER PARTNERSHIP: RENEWABLE ENERGY CERTIFICATES*

U.S. Environmental Protection Agency

WHAT ARE RECS?

RECs represent the environmental and other non-power attributes of renewable electricity generation and are a component of all renewable electricity products. RECs are measured in single megawatt-hour increments and are created at the point of electric generation. Buyers can select RECs based on the generation resource (e.g., wind, solar, geothermal), when the generation occurred, as well as the location of the renewable generator.

RECs provide key information about the generation of renewable electricity delivered to the utility grid. Since RECs represent only the environmental or non-power attributes of renewable electricity generation, they are not subject to electricity delivery constraints. The information conveyed by a REC allows buyers to make specific environmental claims about how their electricity is produced. RECs usually include the following primary attributes and information:

* This is an edited, reformatted and augmented version of the U.S. Environmental Protection Agency, dated July 2008.

- The type of renewable resource producing the electricity
- The vintage of the REC (i.e., the date when it was created)
- The vintage of the renewable generator, or the date when the generator was built
- The renewable generator's location
- The RECs eligibility for certification or renewable portfolio compliance
- The renewable generation's associated greenhouse gas emissions (if any)

RECs are increasingly seen as the "currency" of renewable electricity and green power markets. They can be bought and sold between multiple parties, and they allow their owners to claim that renewable electricity was produced to meet the electricity demand they create.

A REC represents and conveys the environmental and other non-power attributes of one megawatt-hour of renewable electricity generation.

Increasingly, federal, state and local governments are also using RECs as a credible means to meet environmental goals for renewable energy generation. For example, most states allow utilities to use RECs to meet mandated state renewable portfolio standards. State renewable portfolio standards require that a percentage of a utility's electricity generation comes from renewable resources. Increasingly, individuals and organizations are also buying RECs to satisfy a number of other environmental and non-environmental goals:

- Avoid the carbon dioxide (CO_2) emissions associated with conventional electricity use
- Reduce some types of air pollution
- Hedge against future electricity price increases for onsite and some utility products
- Serve as a brand differentiator
- Generate customer, investor, or stakeholder loyalty and employee pride
- Create positive publicity and enhance public image
- Demonstrate civic leadership

How Do RECs Work?

To understand how RECs work, it is helpful to understand how electricity is delivered across the utility grid, as well as what makes renewable electricity generation attractive to individuals and organizational buyers.

Within the United States, electricity demand is met by various types of generation technologies and fuel resources. These electricity generators feed electrons onto the utility grid for delivery to consumers through a complex network of wires and distribution infrastructure. Because the electrons produced from the different technologies and fuel resources are physically the same, it is impossible for individuals or organizations to know what type of generation technology or resource produced the electricity that reaches their particular facility.

RECs help address the issue that the electricity or electrons a consumer receives from their utility does not identify how the electricity was generated. RECs were created to help convey the attributes of electricity generated from renewable resources to buyers. Analogous to the utility delivering the physical electricity through wires, RECs serve as the means to deliver the environmental and non-power attributes of renewable electricity generation to buyers – separate from the physical electricity. (See Figure 1.) All renewable electricity generation can be viewed as having two separate parts:

1. The commodity electricity or electrons
2. The environmental and other non-power attributes of generation represented by a REC

Because RECs are monitored and verified, individuals and organizational buyers can buy RECs and be confident that electricity generated on their behalf was done so with renewable energy resources.

Why Are RECs an Important Option for Consumers?

Individuals and organizations may find that buying RECs separately from their electricity service is a useful way to tap into green power markets. More than half of U.S. electricity customers have an option to purchase some type of green power product from a retail electricity provider, but the rest do not have that choice. RECs provide buyers the option to select renewable resources to meet their electricity demand.

For example, in states that have restructured electricity markets, consumers can often buy green power products by switching electric service providers if their current provider does not offer a green power product. If consumers do not wish to switch electric service providers they also have the option to buy RECs separately from their existing electricity service. This approach allows buyers to avoid the environmental impacts of their electricity, since the REC represents a specific amount of avoided greenhouse gas emissions.

RECs can also be purchased in situations where a utility green power product is available, but this utility product does not have the desired environmental characteristics, resource base, or price. RECs provide more choices and more competitive prices because they are not constrained by where they are created or by transmission bottlenecks.

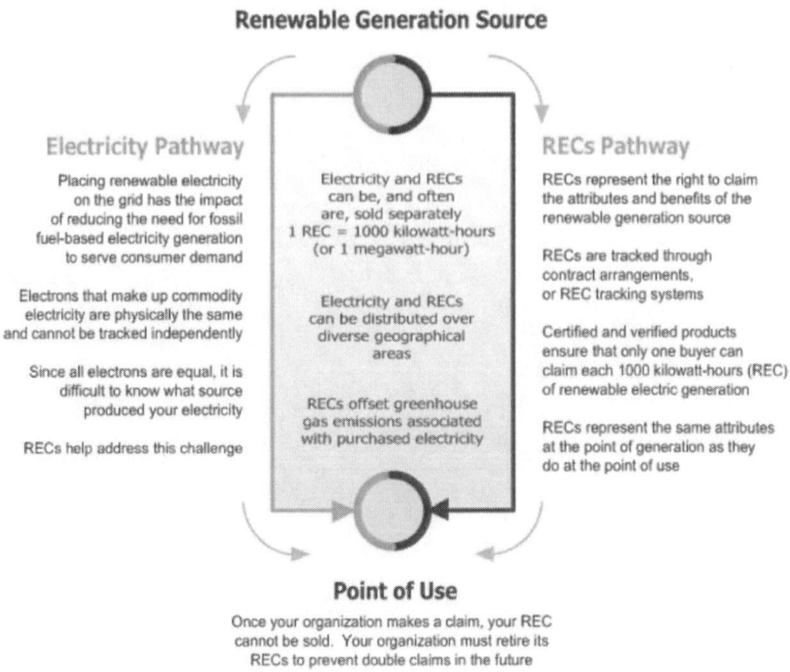

Figure 1.

WHO OWNS A REC?

A REC can be bought and sold between buyers and sellers from its generation to its final point of application or use at a facility. Typically, regional tracking systems register RECs in order to keep track of how much renewable electricity was produced.[1] Tracking systems assign each REC a unique number, which matches the REC to its current owner. Alternatively, buyers can use third-party audits to confirm the contractual chain of ownership between multiple parties. REC contracts typically include a statement, or attestation, from the seller that the RECs have not been sold to, and cannot be claimed by, another party.

WHAT IS THE DIFFERENCE BETWEEN RENEWABLE ELECTRICITY AND GREEN POWER?

Renewable electricity is produced from resources that do not deplete when their energy is harnessed, such as sunlight, wind, waves, water flow, biological processes such as anaerobic digestion (e.g., landfill gas), and geothermal energy. Renewable electricity resources are distinct from fossil and nuclear fuels, which are also used to generate electricity.

EPA defines green power as a subset of renewable electricity and represents those renewable resources and technologies that provide the highest environmental benefit. Green power is renewable electricity produced from solar, wind, geothermal, biogas, biomass, and low-impact small hydroelectric resources. Definitions for renewable energy can vary and may include resources that are acknowledged or perceived to have environmental impacts, such as land use and fisheries impacts of large hydro dams.

RECs are a credible and easy way to keep track of who can claim the environmental attributes of renewable electricity generation through electronic tracking systems. Because RECs are carefully counted, tracked and associated with an owner, no two buyers can legally claim the same environmental benefits of the renewable electricity generation.

Once a buyer makes an environmental claim based on a REC, the buyer can no longer sell the REC and the REC is considered permanently "retired." Buyers can also have their RECs retired in their name by their supplier to ensure that no other entity can lay claim to the same environmental benefits.

Consumers or organizations with onsite renewable electricity systems should be clear about who owns the RECs produced by the onsite system. If the onsite system owner wants to make an environmental claim about the use of renewable electricity from the onsite system, they should ensure that they have and retain ownership of the RECs produced by the onsite renewable electricity system. If the onsite system owner uses the system-produced electricity, but sells the RECs to another party, they are no longer using green power and cannot make a claim to be doing so.

ARE THERE STANDARDS FOR RECS?

There are a number of third-party organizations in the market that certify RECs. As a best practice, EPA recommends that buyers seek out certified products as a form of buyer protection. Certified RECs should meet national standards for resource content and environmental impact. Certification answers the question "Does this product meet accepted standards for quality?"

WHO SELLS RECS?

RECs and the attributes they represent are an ingredient of all green power products. REC providers—including utilities, REC marketers, and other third-party entities—may sell RECs alone or sell them bundled with electricity. As of 2007, more than 50 percent of utility customers have access to green power bundled products, whereas all customers have access to buying renewable energy certificates.

Buyers can identify green power suppliers using EPA's Green Power Locator tool: www.epa.gov/greenpower/pubs/gplocator.htm

Green Power Product Provider/Source	Geographic Availability	Renewable Energy Certificate	Physical Electricity
REC Marketers	Nationally Available	✓	
Utility Green Pricing Programs	Unrestructured Electricity Markets	✓	✓
Utility Green Marketing Programs	Restructured Electricity Markets	✓	✓
Onsite Renewable Electricity Systems	Any Grid-Connected Point of Use	✓	✓

End Note

[1] Regional tracking systems have been put in place to monitor electricity generation across the United States. A REC is produced for every megawatt-hour of electricity generated from a renewable resource and is assigned a tracking number within the system. The tracking number stays with the REC and is transferred between buyers and sellers until a final owner makes a claim, at which time the REC is considered "retired" in the system.

INDEX

D

E

F

G

H

S

T

U

V

W

Y

Z